The Mathematical Proof
The Method and Logic

Contents

Chapter 1

Mathematical proof

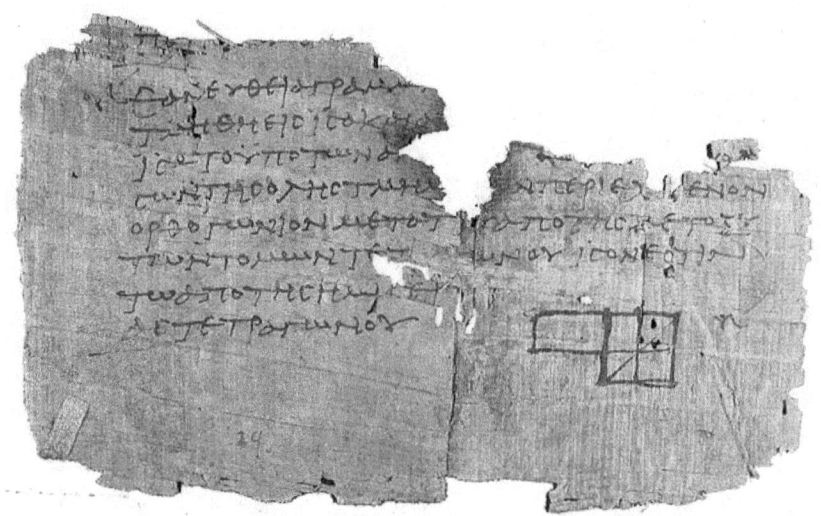

One of the oldest surviving fragments of Euclid's Elements, *a textbook used for millennia to teach proof-writing techniques. The diagram accompanies Book II, Proposition 5.*[1]

In mathematics, a **proof** is a deductive argument for a mathematical statement. In the argument, other previously established statements, such as theorems, can be used. In principle, a proof can be traced back to self-evident or assumed statements, known as axioms.[2][3][4] Proofs are examples of deductive reasoning and are distinguished from inductive or empirical arguments; a proof must demonstrate that a statement is always true (occasionally by listing *all* possible cases and showing that it holds in each), rather than enumerate many confirmatory cases. An unproved proposition that is believed true is known as a conjecture.

Proofs employ logic but usually include some amount of natural language which usually admits some ambiguity. In fact, the vast majority of proofs in written mathematics can be considered as applications of rigorous informal logic. Purely formal proofs, written in symbolic language instead of natural language, are considered in proof theory. The distinction between formal and informal proofs has led to much examination of current and historical mathematical practice, quasi-empiricism in mathematics, and so-called folk mathematics (in both senses of that term). The philosophy of mathematics

1

is concerned with the role of language and logic in proofs, and mathematics as a language.

1.1   History and etymology

See also: History of logic

The word "proof" comes from the Latin *probare* meaning "to test". Related modern words are the English "probe", "probation", and "probability", the Spanish *probar* (to smell or taste, or (lesser use) touch or test),[5] Italian *provare* (to try), and the German *probieren* (to try). The early use of "probity" was in the presentation of legal evidence. A person of authority, such as a nobleman, was said to have probity, whereby the evidence was by his relative authority, which outweighed empirical testimony.[6]

Plausibility arguments using heuristic devices such as pictures and analogies preceded strict mathematical proof.[7] It is likely that the idea of demonstrating a conclusion first arose in connection with geometry, which originally meant the same as "land measurement".[8] The development of mathematical proof is primarily the product of ancient Greek mathematics, and one of its greatest achievements. Thales (624–546 BCE) proved some theorems in geometry. Eudoxus (408–355 BCE) and Theaetetus (417–369 BCE) formulated theorems but did not prove them. Aristotle (384–322 BCE) said definitions should describe the concept being defined in terms of other concepts already known. Mathematical proofs were revolutionized by Euclid (300 BCE), who introduced the axiomatic method still in use today, starting with undefined terms and axioms (propositions regarding the undefined terms assumed to be self-evidently true from the Greek "axios" meaning "something worthy"), and used these to prove theorems using deductive logic. His book, the *Elements*, was read by anyone who was considered educated in the West until the middle of the 20th century.[9] In addition to the familiar theorems of geometry, such as the Pythagorean theorem, the *Elements* includes a proof that the square root of two is irrational and that there are infinitely many prime numbers.

Further advances took place in medieval Islamic mathematics. While earlier Greek proofs were largely geometric demonstrations, the development of arithmetic and algebra by Islamic mathematicians allowed more general proofs that no longer depended on geometry. In the 10th century CE, the Iraqi mathematician Al-Hashimi provided general proofs for numbers (rather than geometric demonstrations) as he considered multiplication, division, etc. for "lines." He used this method to provide a proof of the existence of irrational numbers.[10] An inductive proof for arithmetic sequences was introduced in the *Al-Fakhri* (1000) by Al-Karaji, who used it to prove the binomial theorem and properties of Pascal's triangle. Alhazen also developed the method of proof by contradiction, as the first attempt at proving the Euclidean parallel postulate.[11]

Modern proof theory treats proofs as inductively defined data structures. There is no longer an assumption that axioms are "true" in any sense; this allows for parallel mathematical theories built on alternate sets of axioms (see Axiomatic set theory and Non-Euclidean geometry for examples).

1.2   Nature and purpose

As practiced, a proof is expressed in natural language and is a rigorous argument intended to convince the audience of the truth of a statement. The standard of rigor is not absolute and has varied throughout history. A proof can be presented differently depending on the intended audience. In order to gain acceptance, a proof has to meet communal statements of rigor; an argument considered vague or incomplete may be rejected.

The concept of a proof is formalized in the field of mathematical logic.[12] A formal proof is written in a formal language instead of a natural language. A formal proof is defined as sequence of formulas in a formal language, in which each formula is a logical consequence of preceding formulas. Having a definition of formal proof makes the concept of proof amenable to study. Indeed, the field of proof theory studies formal proofs and their properties, for example, the property that a statement has a formal proof. An application of proof theory is to show that certain undecidable statements are not provable.

The definition of a formal proof is intended to capture the concept of proofs as written in the practice of mathematics. The soundness of this definition amounts to the belief that a published proof can, in principle, be converted into a formal proof. However, outside the field of automated proof assistants, this is rarely done in practice. A classic question in philosophy asks whether mathematical proofs are analytic or synthetic. Kant, who introduced the analytic-synthetic distinction, believed mathematical proofs are synthetic.

Proofs may be viewed as aesthetic objects, admired for their mathematical beauty. The mathematician Paul Erdős was known for describing proofs he found particularly elegant as coming from "The Book", a hypothetical tome containing the most beautiful method(s) of proving each theorem. The book *Proofs from THE BOOK*, published in 2003, is devoted to presenting 32 proofs its editors find particularly pleasing.

1.3 Methods

1.3.1 Direct proof

Main article: Direct proof

In direct proof, the conclusion is established by logically combining the axioms, definitions, and earlier theorems.[13] For example, direct proof can be used to establish that the sum of two even integers is always even:

> Consider two even integers x and y. Since they are even, they can be written as $x = 2a$ and $y = 2b$, respectively, for integers a and b. Then the sum $x + y = 2a + 2b = 2(a+b)$. Therefore $x+y$ has 2 as a factor and, by definition, is even. Hence the sum of any two even integers is even.

This proof uses the definition of even integers, the integer properties of closure under addition and multiplication, and distributivity.

1.3.2 Proof by mathematical induction

Main article: Mathematical induction

Mathematical induction is not a form of inductive reasoning. In proof by mathematical induction, a single "base case" is proved, and an "induction rule" is proved, which establishes that a certain case implies the next case. Applying the induction rule repeatedly, starting from the independently proved base case, proves many, often infinitely many, other cases.[14] Since the base case is true, the infinity of other cases must also be true, even if all of them cannot be proved directly because of their infinite number. A subset of induction is infinite descent. Infinite descent can be used to prove the irrationality of the square root of two.

A common application of proof by mathematical induction is to prove that a property known to hold for one number holds for all natural numbers:[15] Let $\mathbf{N} = \{1,2,3,4,...\}$ be the set of natural numbers, and $P(n)$ be a mathematical statement involving the natural number n belonging to $\mathbf{N}$ such that

- (i) $P(1)$ is true, i.e., $P(n)$ is true for $n = 1$.
- (ii) $P(n+1)$ is true whenever $P(n)$ is true, i.e., $P(n)$ is true implies that $P(n+1)$ is true.
- Then $P(n)$ is true for all natural numbers n.

For example, we can prove by induction that all integers of the form $2n + 1$ are odd:

> (i) For $n = 1$, $2n + 1 = 2(1) + 1 = 3$, and 3 is odd. Thus $P(1)$ is true.
>
> (ii) For $2n + 1$ for some n, $2(n+1) + 1 = (2n+1) + 2$. If $2n + 1$ is odd, then $(2n+1) + 2$ must also be odd, because adding 2 to an odd number results in an odd number. So $P(n+1)$ is true if $P(n)$ is true.
>
> **Thus** $2n + 1$ is odd, for all natural numbers n.

It is common for the phrase "proof by induction" to be used for a "proof by mathematical induction".[16]

1.3.3 Proof by contraposition

Main article: Contraposition

Proof by contraposition infers the conclusion "if p then q" from the premise "if *not q* then *not p*". The statement "if *not q* then *not p*" is called the contrapositive of the statement "if p then q". For example, contraposition can be used to establish that, given an integer x, if x^2 is even, then x is even:

> Suppose x is not even. Then x is odd. The product of two odd numbers is odd, hence $x^2 = x \cdot x$ is odd. Thus x^2 is not even.

1.3.4 Proof by contradiction

Main article: Proof by contradiction

In proof by contradiction (also known as *reductio ad absurdum*, Latin for "by reduction to the absurd"), it is shown that if some statement were true, a logical contradiction occurs, hence the statement must be false. A famous example of proof by contradiction shows that $\sqrt{2}$ is an irrational number:

> Suppose that $\sqrt{2}$ were a rational number, so by definition $\sqrt{2} = \frac{a}{b}$ where a and b are non-zero integers with no common factor. Thus, $b\sqrt{2} = a$. Squaring both sides yields $2b^2 = a^2$. Since 2 divides the left hand side, 2 must also divide the right hand side (as they are equal and both integers). So a^2 is even, which implies that a must also be even. So we can write $a = 2c$, where c is also an integer. Substitution into the original equation yields $2b^2 = (2c)^2 = 4c^2$. Dividing both sides by 2 yields $b^2 = 2c^2$. But then, by the same argument as before, 2 divides b^2, so b must be even. However, if a and b are both even, they share a factor, namely 2. This contradicts our assumption, so we are forced to conclude that $\sqrt{2}$ is an irrational number.

1.3.5 Proof by construction

Main article: Proof by construction

Proof by construction, or proof by example, is the construction of a concrete example with a property to show that something having that property exists. Joseph Liouville, for instance, proved the existence of transcendental numbers by constructing an explicit example. It can also be used to construct a counterexample to disprove a proposition that all elements have a certain property.

1.3.6 Proof by exhaustion

Main article: Proof by exhaustion

In proof by exhaustion, the conclusion is established by dividing it into a finite number of cases and proving each one separately. The number of cases sometimes can become very large. For example, the first proof of the four color theorem was a proof by exhaustion with 1,936 cases. This proof was controversial because the majority of the cases were checked by a computer program, not by hand. The shortest known proof of the four color theorem as of 2011 still has over 600 cases.

1.3.7 Probabilistic proof

Main article: Probabilistic method

A probabilistic proof is one in which an example is shown to exist, with certainty, by using methods of probability theory. Probabilistic proof, like proof by construction, is one of many ways to show existence theorems.

This is not to be confused with an argument that a theorem is 'probably' true, a 'plausibility argument'. The work on the Collatz conjecture shows how far plausibility is from genuine proof.[17]

1.3.8 Combinatorial proof

Main article: Combinatorial proof

A combinatorial proof establishes the equivalence of different expressions by showing that they count the same object in different ways. Often a bijection between two sets is used to show that the expressions for their two sizes are equal. Alternatively, a double counting argument provides two different expressions for the size of a single set, again showing that the two expressions are equal.

1.3.9 Nonconstructive proof

Main article: Nonconstructive proof

A nonconstructive proof establishes that a mathematical object with a certain property exists without explaining how such an object can be found. Often, this takes the form of a proof by contradiction in which the nonexistence of the object is proved to be impossible. In contrast, a constructive proof establishes that a particular object exists by providing a method of finding it. A famous example of a nonconstructive proof shows that there exist two irrational numbers a and b such that a^b is a rational number:

Either $\sqrt{2}^{\sqrt{2}}$ is a rational number and we are done (take $a = b = \sqrt{2}$), or $\sqrt{2}^{\sqrt{2}}$ is irrational so we can write $a = \sqrt{2}^{\sqrt{2}}$ and $b = \sqrt{2}$. This then gives $\left(\sqrt{2}^{\sqrt{2}} \right)^{\sqrt{2}} = \sqrt{2}^2 = 2$, which is thus a rational of the form a^b.

1.3.10 Statistical proofs in pure mathematics

Main article: Statistical proof

The expression "statistical proof" may be used technically or colloquially in areas of pure mathematics, such as involving cryptography, chaotic series, and probabilistic or analytic number theory.[18][19][20] It is less commonly used to refer to a mathematical proof in the branch of mathematics known as mathematical statistics. See also "Statistical proof using data" section below.

1.3.11 Computer-assisted proofs

Main article: Computer-assisted proof

Until the twentieth century it was assumed that any proof could, in principle, be checked by a competent mathematician to confirm its validity.[7] However, computers are now used both to prove theorems and to carry out calculations that are too long for any human or team of humans to check; the first proof of the four color theorem is an example of a computer-assisted proof. Some mathematicians are concerned that the possibility of an error in a computer program or a run-time error in its calculations calls the validity of such computer-assisted proofs into question. In practice, the chances of an error invalidating a computer-assisted proof can be reduced by incorporating redundancy and self-checks into calculations, and by developing multiple independent approaches and programs. Errors can never be completely ruled out in case of verification of a proof by humans either, especially if the proof contains natural language and requires deep mathematical insight.

1.4 Undecidable statements

A statement that is neither provable nor disprovable from a set of axioms is called undecidable (from those axioms). One example is the parallel postulate, which is neither provable nor refutable from the remaining axioms of Euclidean geometry.

Mathematicians have shown there are many statements that are neither provable nor disprovable in Zermelo-Fraenkel set theory with the axiom of choice (ZFC), the standard system of set theory in mathematics (assuming that ZFC is consistent); see list of statements undecidable in ZFC.

Gödel's (first) incompleteness theorem shows that many axiom systems of mathematical interest will have undecidable statements.

1.5 Heuristic mathematics and experimental mathematics

Main article: Experimental mathematics

While early mathematicians such as Eudoxus of Cnidus did not use proofs, from Euclid to the foundational mathematics developments of the late 19th and 20th centuries, proofs were an essential part of mathematics.[21] With the increase in computing power in the 1960s, significant work began to be done investigating mathematical objects outside of the proof-theorem framework,[22] in experimental mathematics. Early pioneers of these methods intended the work ultimately to be embedded in a classical proof-theorem framework, e.g. the early development of fractal geometry,[23] which was ultimately so embedded.

1.6 Related concepts

1.6.1 Visual proof

Although not a formal proof, a visual demonstration of a mathematical theorem is sometimes called a "proof without words". The left-hand picture below is an example of a historic visual proof of the Pythagorean theorem in the case of the (3,4,5) triangle.

- Visual proof for the (3, 4, 5) triangle as in the Chou Pei Suan Ching 500–200 BC.
- Animated visual proof for the Pythagorean theorem by rearrangement.
- A second animated proof of the Pythagorean theorem.

Some illusory visual proofs, such as the missing square puzzle, can be constructed in a way which appear to prove a supposed mathematical fact but only do so under the presence of tiny errors (for example, supposedly straight lines which actually bend slightly) which are unnoticeable until the entire picture is closely examined, with lengths and angles precisely measured or calculated.

1.6.2 Elementary proof

Main article: Elementary proof

An elementary proof is a proof which only uses basic techniques. More specifically, the term is used in number theory to refer to proofs that make no use of complex analysis. For some time it was thought that certain theorems, like the prime number theorem, could only be proved using "higher" mathematics. However, over time, many of these results have been reproved using only elementary techniques.

1.6.3 Two-column proof

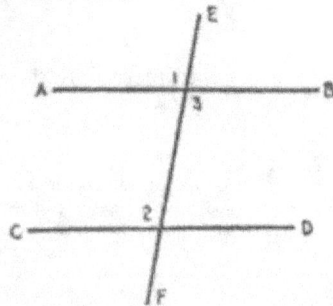

PROPOSITION XIX. THEOREM

106. *If two parallel lines are cut by a transversal, the corresponding angles are equal.*

[Converse of Prop. XIV.]

Given parallel lines AB and CD and the cor. $\angle$ 1 and 2.

To prove $\angle 1 = \angle 2$.

Proof

STATEMENTS	REASONS
$\angle 1 = \angle 3$.	Vertical $\angle$ are equal.
$\angle 2 = \angle 3$.	Alt. int. $\angle$ of $\parallel$ lines are equal.
$\therefore \angle 1 = \angle 2$.	Things equal to the same thing
Q. E. D.	are equal to each other.

A two-column proof published in 1913

A particular way of organising a proof using two parallel columns is often used in elementary geometry classes in the United States.[24] The proof is written as a series of lines in two columns. In each line, the left-hand column contains a proposition, while the right-hand column contains a brief explanation of how the corresponding proposition in the left-hand column is either an axiom, a hypothesis, or can be logically derived from previous propositions. The left-hand column is typically headed "Statements" and the right-hand column is typically headed "Reasons".[25]

1.6.4   Colloquial use of "mathematical proof"

The expression "mathematical proof" is used by lay people to refer to using mathematical methods or arguing with mathematical objects, such as numbers, to demonstrate something about everyday life, or when data used in an argument is numerical. It is sometimes also used to mean a "statistical proof" (below), especially when used to argue from data.

1.6.5   Statistical proof using data

Main article: Statistical proof

"Statistical proof" from data refers to the application of statistics, data analysis, or Bayesian analysis to infer propositions regarding the probability of data. While *using* mathematical proof to establish theorems in statistics, it is usually not a mathematical proof in that the *assumptions* from which probability statements are derived require empirical evidence from outside mathematics to verify. In physics, in addition to statistical methods, "statistical proof" can refer to the specialized *mathematical methods of physics* applied to analyze data in a particle physics experiment or observational study in cosmology. "Statistical proof" may also refer to raw data or a convincing diagram involving data, such as scatter plots, when the data or diagram is adequately convincing without further analysis.

1.6.6   Inductive logic proofs and Bayesian analysis

Main articles: Inductive logic and Bayesian analysis

Proofs using inductive logic, while considered mathematical in nature, seek to establish propositions with a degree of certainty, which acts in a similar manner to probability, and may be less than one certainty. Bayesian analysis establishes assertions as to the degree of a person's subjective belief. Inductive logic should not be confused with mathematical induction.

1.6.7   Proofs as mental objects

Main articles: Psychologism and Language of thought

Psychologism views mathematical proofs as psychological or mental objects. Mathematician philosophers, such as Leibniz, Frege, and Carnap have attempted to develop a semantics for what they considered to be the language of thought, whereby standards of mathematical proof might be applied to empirical science.

1.6.8   Influence of mathematical proof methods outside mathematics

Philosopher-mathematicians such as Spinoza have attempted to formulate philosophical arguments in an axiomatic manner, whereby mathematical proof standards could be applied to argumentation in general philosophy. Other mathematician-philosophers have tried to use standards of mathematical proof and reason, without empiricism, to arrive at statements outside of mathematics, but having the certainty of propositions deduced in a mathematical proof, such as Descarte's *cogito* argument.

1.7 Ending a proof

Main article: Q.E.D.

Sometimes, the abbreviation *"Q.E.D."* is written to indicate the end of a proof. This abbreviation stands for *"Quod Erat Demonstrandum"*, which is Latin for *"that which was to be demonstrated"*. A more common alternative is to use a square or a rectangle, such as □ or ▮, known as a "tombstone" or "halmos" after its eponym Paul Halmos. Often, "which was to be shown" is verbally stated when writing "QED", "□", or "▮" in an oral presentation on a board.

1.8 See also

- Automated theorem proving
- Invalid proof
- List of incomplete proofs
- List of long proofs
- List of mathematical proofs
- Nonconstructive proof
- Proof by intimidation
- Termination analysis
- *What the Tortoise Said to Achilles*

1.9 References

[1] Bill Casselman. "One of the Oldest Extant Diagrams from Euclid". University of British Columbia. Retrieved 2008-09-26.

[2] Clapham, C. and Nicholson, JN. *The Concise Oxford Dictionary of Mathematics, Fourth edition*. A statement whose truth is either to be taken as self-evident or to be assumed. Certain areas of mathematics involve choosing a set of axioms and discovering what results can be derived from them, providing proofs for the theorems that are obtained.

[3] Cupillari, Antonella. *The Nuts and Bolts of Proofs*. Academic Press, 2001. Page 3.

[4] Gossett, Eric. *Discrete Mathematics with Proof*. John Wiley and Sons, 2009. Definition 3.1 page 86. ISBN 0-470-45793-7

[5] New Shorter Oxford English Dictionary, 1993, OUP, Oxford.

[6] The Emergence of Probability, Ian Hacking

[7] The History and Concept of Mathematical Proof, Steven G. Krantz. 1. February 5, 2007

[8] Kneale, p. 2

[9] Howard Eves, *An Introduction to the History of Mathematics*, Saunders, 1990, ISBN 0-03-029558-0 p. 141: "No work, except The Bible, has been more widely used...."

[10] Matvievskaya, Galina (1987), "The Theory of Quadratic Irrationals in Medieval Oriental Mathematics", *Annals of the New York Academy of Sciences* **500**: 253–277 [260], doi:10.1111/j.1749-6632.1987.tb37206.x

[11] Eder, Michelle (2000), *Views of Euclid's Parallel Postulate in Ancient Greece and in Medieval Islam*, Rutgers University, retrieved 2008-01-23

[12] Buss, 1997, p. 3

[13] Cupillari, page 20.

[14] Cupillari, page 46.

[15] Examples of simple proofs by mathematical induction for all natural numbers

[16] Proof by induction, University of Warwick Glossary of Mathematical Terminology

[17] While most mathematicians do not think that probabilistic evidence ever counts as a genuine mathematical proof, a few mathematicians and philosophers have argued that at least some types of probabilistic evidence (such as Rabin's probabilistic algorithm for testing primality) are as good as genuine mathematical proofs. See, for example, Davis, Philip J. (1972), "Fidelity in Mathematical Discourse: Is One and One Really Two?" *American Mathematical Monthly* 79:252-63. Fallis, Don (1997), "The Epistemic Status of Probabilistic Proof." *Journal of Philosophy* 94:165-86.

[18] "in number theory and commutative algebra... in particular the statistical proof of the lemma."

[19] "Whether constant π (i.e., pi) is normal is a confusing problem without any strict theoretical demonstration except for some **statistical** proof"" (Derogatory use.)

[20] "these observations suggest a statistical proof of Goldbach's conjecture with very quickly vanishing probability of failure for large E"

[21] "*What to do with the pictures? Two thoughts surfaced: the first was that they were unpublishable in the standard way, there were no theorems only very suggestive pictures. They furnished convincing evidence for many conjectures and lures to further exploration, but theorems were coins of the realm ant the conventions of that day dictated that journals only published theorems*", David Mumford, Caroline Series and David Wright, Indra's Pearls, 2002

[22] "*Mandelbrot, working at the IBM Research Laboratory, did some computer simulations for these sets on the reasonable assumption that, if you wanted to prove something, it might be helpful to know the answer ahead of time.*"A Note on the History of Fractals,

[23] "*... brought home again to Benoit [Mandelbrot] that there was a 'mathematics of the eye', that visualization of a problem was as valid a method as any for finding a solution. Amazingly, he found himself alone with this conjecture. The teaching of mathematics in France was dominated by a handful of dogmatic mathematicians hiding behind the pseudonym 'Bourbaki'...* ", Introducing Fractal Geometry, Nigel Lesmoir-Gordon

[24] Patricio G. Herbst, Establishing a Custom of Proving in American School Geometry: Evolution of the Two-Column Proof in the Early Twentieth Century, Educational Studies in Mathematics, Vol. 49, No. 3 (2002), pp. 283-312,

[25] Introduction to the Two-Column Proof, Carol Fisher

1.10   Sources

- Pólya, G. (1954), *Mathematics and Plausible Reasoning*, Princeton University Press.

- Fallis, Don (2002), "What Do Mathematicians Want? Probabilistic Proofs and the Epistemic Goals of Mathematicians", *Logique et Analyse* **45**: 373–388.

- Franklin, J.; Daoud, A. (2011), *Proof in Mathematics: An Introduction*, Kew Books, ISBN 0-646-54509-4.

- Solow, D. (2004), *How to Read and Do Proofs: An Introduction to Mathematical Thought Processes*, Wiley, ISBN 0-471-68058-3.

- Velleman, D. (2006), *How to Prove It: A Structured Approach*, Cambridge University Press, ISBN 0-521-67599-5.

1.11   External links

- Proofs in Mathematics: Simple, Charming and Fallacious

- A lesson about proofs, in a course from Wikiversity

Chapter 2

Philosophy of mathematics

The **philosophy of mathematics** is the branch of philosophy that studies the philosophical assumptions, foundations, and implications of mathematics. The aim of the philosophy of mathematics is to provide an account of the nature and methodology of mathematics and to understand the place of mathematics in people's lives. The logical and structural nature of mathematics itself makes this study both broad and unique among its philosophical counterparts.

The terms *philosophy of mathematics* and *mathematical philosophy* are frequently used as synonyms.[1] The latter, however, may be used to refer to several other areas of study. One refers to a project of formalizing a philosophical subject matter, say, aesthetics, ethics, logic, metaphysics, or theology, in a purportedly more exact and rigorous form, as for example the labors of scholastic theologians, or the systematic aims of Leibniz and Spinoza. Another refers to the working philosophy of an individual practitioner or a like-minded community of practicing mathematicians. Additionally, some understand the term "mathematical philosophy" to be an allusion to the approach to the foundations of mathematics taken by Bertrand Russell in his books *The Principles of Mathematics* and *Introduction to Mathematical Philosophy*.

2.1 Recurrent themes

Recurrent themes include:

- What is the role of Mankind in developing mathematics?
- What are the sources of mathematical subject matter?
- What is the ontological status of mathematical entities?
- What does it mean to refer to a mathematical object?
- What is the character of a mathematical proposition?
- What is the relation between logic and mathematics?
- What is the role of hermeneutics in mathematics?
- What kinds of inquiry play a role in mathematics?
- What are the objectives of mathematical inquiry?
- What gives mathematics its hold on experience?
- What are the human traits behind mathematics?
- What is mathematical beauty?
- What is the source and nature of mathematical truth?
- What is the relationship between the abstract world of mathematics and the material universe?

2.2   History

The origin of mathematics is subject to argument. Whether the birth of mathematics was a random happening or induced by necessity duly contingent upon other subjects, say for example physics, is still a matter of prolific debates.

Many thinkers have contributed their ideas concerning the nature of mathematics. Today, some philosophers of mathematics aim to give accounts of this form of inquiry and its products as they stand, while others emphasize a role for themselves that goes beyond simple interpretation to critical analysis. There are traditions of mathematical philosophy in both Western philosophy and Eastern philosophy. Western philosophies of mathematics go as far back as Plato, who studied the ontological status of mathematical objects, and Aristotle, who studied logic and issues related to infinity (actual versus potential).

Greek philosophy on mathematics was strongly influenced by their study of geometry. For example, at one time, the Greeks held the opinion that 1 (one) was not a number, but rather a unit of arbitrary length. A number was defined as a multitude. Therefore 3, for example, represented a certain multitude of units, and was thus not "truly" a number. At another point, a similar argument was made that 2 was not a number but a fundamental notion of a pair. These views come from the heavily geometric straight-edge-and-compass viewpoint of the Greeks: just as lines drawn in a geometric problem are measured in proportion to the first arbitrarily drawn line, so too are the numbers on a number line measured in proportion to the arbitrary first "number" or "one".

These earlier Greek ideas of numbers were later upended by the discovery of the irrationality of the square root of two. Hippasus, a disciple of Pythagoras, showed that the diagonal of a unit square was incommensurable with its (unit-length) edge: in other words he proved there was no existing (rational) number that accurately depicts the proportion of the diagonal of the unit square to its edge. This caused a significant re-evaluation of Greek philosophy of mathematics. According to legend, fellow Pythagoreans were so traumatized by this discovery that they murdered Hippasus to stop him from spreading his heretical idea. Simon Stevin was one of the first in Europe to challenge Greek ideas in the 16th century. Beginning with Leibniz, the focus shifted strongly to the relationship between mathematics and logic. This perspective dominated the philosophy of mathematics through the time of Frege and of Russell, but was brought into question by developments in the late 19th and early 20th centuries.

2.2.1   20th century

A perennial issue in the philosophy of mathematics concerns the relationship between logic and mathematics at their joint foundations. While 20th century philosophers continued to ask the questions mentioned at the outset of this article, the philosophy of mathematics in the 20th century was characterized by a predominant interest in formal logic, set theory, and foundational issues.

It is a profound puzzle that on the one hand mathematical truths seem to have a compelling inevitability, but on the other hand the source of their "truthfulness" remains elusive. Investigations into this issue are known as the foundations of mathematics program.

At the start of the 20th century, philosophers of mathematics were already beginning to divide into various schools of thought about all these questions, broadly distinguished by their pictures of mathematical epistemology and ontology. Three schools, formalism, intuitionism, and logicism, emerged at this time, partly in response to the increasingly widespread worry that mathematics as it stood, and analysis in particular, did not live up to the standards of certainty and rigor that had been taken for granted. Each school addressed the issues that came to the fore at that time, either attempting to resolve them or claiming that mathematics is not entitled to its status as our most trusted knowledge.

Surprising and counter-intuitive developments in formal logic and set theory early in the 20th century led to new questions concerning what was traditionally called the *foundations of mathematics*. As the century unfolded, the initial focus of concern expanded to an open exploration of the fundamental axioms of mathematics, the axiomatic approach having been taken for granted since the time of Euclid around 300 BCE as the natural basis for mathematics. Notions of axiom, proposition and proof, as well as the notion of a proposition being true of a mathematical object (see Assignment (mathematical logic)), were formalized, allowing them to be treated mathematically. The Zermelo–Fraenkel axioms for set theory were formulated which provided a conceptual framework in which much mathematical discourse would be interpreted. In mathematics, as in physics, new and unexpected ideas had arisen and significant changes were coming. With

Gödel numbering, propositions could be interpreted as referring to themselves or other propositions, enabling inquiry into the consistency of mathematical theories. This reflective critique in which the theory under review "becomes itself the object of a mathematical study" led Hilbert to call such study *metamathematics* or *proof theory*.[2]

At the middle of the century, a new mathematical theory was created by Samuel Eilenberg and Saunders Mac Lane, known as category theory, and it became a new contender for the natural language of mathematical thinking.[3] As the 20th century progressed, however, philosophical opinions diverged as to just how well-founded were the questions about foundations that were raised at the century's beginning. Hilary Putnam summed up one common view of the situation in the last third of the century by saying:

> When philosophy discovers something wrong with science, sometimes science has to be changed— Russell's paradox comes to mind, as does Berkeley's attack on the actual infinitesimal—but more often it is philosophy that has to be changed. I do not think that the difficulties that philosophy finds with classical mathematics today are genuine difficulties; and I think that the philosophical interpretations of mathematics that we are being offered on every hand are wrong, and that "philosophical interpretation" is just what mathematics doesn't need.[4]:169–170

Philosophy of mathematics today proceeds along several different lines of inquiry, by philosophers of mathematics, logicians, and mathematicians, and there are many schools of thought on the subject. The schools are addressed separately in the next section, and their assumptions explained.

2.3 Major themes

2.3.1 Mathematical realism

Mathematical realism, like realism in general, holds that mathematical entities exist independently of the human mind. Thus humans do not invent mathematics, but rather discover it, and any other intelligent beings in the universe would presumably do the same. In this point of view, there is really one sort of mathematics that can be discovered; triangles, for example, are real entities, not the creations of the human mind.

Many working mathematicians have been mathematical realists; they see themselves as discoverers of naturally occurring objects. Examples include Paul Erdős and Kurt Gödel. Gödel believed in an objective mathematical reality that could be perceived in a manner analogous to sense perception. Certain principles (e.g., for any two objects, there is a collection of objects consisting of precisely those two objects) could be directly seen to be true, but the continuum hypothesis conjecture might prove undecidable just on the basis of such principles. Gödel suggested that quasi-empirical methodology could be used to provide sufficient evidence to be able to reasonably assume such a conjecture.

Within realism, there are distinctions depending on what sort of existence one takes mathematical entities to have, and how we know about them. Major forms of mathematical realism include Platonism and empiricism.

2.3.2 Mathematical anti-realism

Mathematical anti-realism generally holds that mathematical statements have truth-values, but that they do not do so by corresponding to a special realm of immaterial or non-empirical entities. Major forms of mathematical anti-realism include Formalism and Fictionalism.

2.4 Contemporary schools of thought

2.4.1 Platonism

Mathematical Platonism is the form of realism that suggests that mathematical entities are abstract, have no spatiotemporal or causal properties, and are eternal and unchanging. This is often claimed to be the view most people have of numbers.

The term *Platonism* is used because such a view is seen to parallel Plato's Theory of Forms and a "World of Ideas" (Greek: *eidos* (εἶδος)) described in Plato's allegory of the cave: the everyday world can only imperfectly approximate an unchanging, ultimate reality. Both *Plato's cave* and *Platonism* have meaningful, not just superficial connections, because Plato's ideas were preceded and probably influenced by the hugely popular *Pythagoreans* of ancient Greece, who believed that the world was, quite literally, generated by numbers.

A major question considered in mathematical platonism is this: precisely where and how do the mathematical entities exist, and how do we know about them? Is there a world, completely separate from our physical one, that is occupied by the mathematical entities? How can we gain access to this separate world and discover truths about the entities? One answer might be the Ultimate Ensemble, which is a theory that postulates all structures that exist mathematically also exist physically in their own universe.

Plato spoke of mathematics by:

> How do you mean?
> I mean, as I was saying, that arithmetic has a very great and elevating effect, compelling the soul to reason about abstract number, and rebelling against the introduction of visible or tangible objects into the argument. You know how steadily the masters of the art repel and ridicule any one who attempts to divide absolute unity when he is calculating, and if you divide, they multiply, taking care that one shall continue one and not become lost in fractions.
> That is very true.
> Now, suppose a person were to say to them: O my friends, what are these wonderful numbers about which you are reasoning, in which, as you say, there is a unity such as you demand, and each unit is equal, invariable, indivisible, --what would they answer?
> —Plato, Chapter 7. "The Republic" (Jowett translation).

In context, chapter 8, of H.D.P. Lee's translation, reports the education of a philosopher contains five mathematical disciplines:

1. mathematics;

2. arithmetic, written in unit fraction "parts" using theoretical unities and abstract numbers;

3. plane geometry and solid geometry also considered the line to be segmented into rational and irrational unit "parts";

4. astronomy

5. harmonics

Translators of the works of Plato rebelled against practical versions of his culture's practical mathematics. However, Plato himself and Greeks had copied 1,500 older Egyptian fraction abstract unities, one being a hekat unity scaled to (64/64) in the Akhmim Wooden Tablet, thereby not getting lost in fractions.

Gödel's Platonism postulates a special kind of mathematical intuition that lets us perceive mathematical objects directly. (This view bears resemblances to many things Husserl said about mathematics, and supports Kant's idea that mathematics is synthetic *a priori*.) Davis and Hersh have suggested in their book *The Mathematical Experience* that most mathematicians act as though they are Platonists, even though, if pressed to defend the position carefully, they may retreat to formalism (see below).

Some mathematicians hold opinions that amount to more nuanced versions of Platonism.

Full-blooded Platonism is a modern variation of Platonism, which is in reaction to the fact that different sets of mathematical entities can be proven to exist depending on the axioms and inference rules employed (for instance, the law of the excluded middle, and the axiom of choice). It holds that all mathematical entities exist, however they may be provable, even if they cannot all be derived from a single consistent set of axioms.

2.4.2 Empiricism

Empiricism is a form of realism that denies that mathematics can be known *a priori* at all. It says that we discover mathematical facts by empirical research, just like facts in any of the other sciences. It is not one of the classical three positions advocated in the early 20th century, but primarily arose in the middle of the century. However, an important early proponent of a view like this was John Stuart Mill. Mill's view was widely criticized, because, according to critics, it makes statements like "2 + 2 = 4" come out as uncertain, contingent truths, which we can only learn by observing instances of two pairs coming together and forming a quartet.

Contemporary mathematical empiricism, formulated by Quine and Putnam, is primarily supported by the indispensability argument: mathematics is indispensable to all empirical sciences, and if we want to believe in the reality of the phenomena described by the sciences, we ought also believe in the reality of those entities required for this description. That is, since physics needs to talk about electrons to say why light bulbs behave as they do, then electrons must exist. Since physics needs to talk about numbers in offering any of its explanations, then numbers must exist. In keeping with Quine and Putnam's overall philosophies, this is a naturalistic argument. It argues for the existence of mathematical entities as the best explanation for experience, thus stripping mathematics of being distinct from the other sciences.

Putnam strongly rejected the term "Platonist" as implying an over-specific ontology that was not necessary to mathematical practice in any real sense. He advocated a form of "pure realism" that rejected mystical notions of truth and accepted much quasi-empiricism in mathematics. Putnam was involved in coining the term "pure realism" (see below).

The most important criticism of empirical views of mathematics is approximately the same as that raised against Mill. If mathematics is just as empirical as the other sciences, then this suggests that its results are just as fallible as theirs, and just as contingent. In Mill's case the empirical justification comes directly, while in Quine's case it comes indirectly, through the coherence of our scientific theory as a whole, i.e. consilience after E.O. Wilson. Quine suggests that mathematics seems completely certain because the role it plays in our web of belief is incredibly central, and that it would be extremely difficult for us to revise it, though not impossible.

For a philosophy of mathematics that attempts to overcome some of the shortcomings of Quine and Gödel's approaches by taking aspects of each see Penelope Maddy's *Realism in Mathematics*. Another example of a realist theory is the embodied mind theory (below). For a modern revision of mathematical empiricism see New Empiricism (below).

For experimental evidence suggesting that human infants can do elementary arithmetic, see Brian Butterworth.

2.4.3 Mathematical monism

Max Tegmark's mathematical universe hypothesis goes further than full-blooded Platonism in asserting that not only do all mathematical objects exist, but nothing else does. Tegmark's sole postulate is: *All structures that exist mathematically also exist physically*. That is, in the sense that "in those [worlds] complex enough to contain self-aware substructures [they] will subjectively perceive themselves as existing in a physically 'real' world".[5][6]

2.4.4 Logicism

Logicism is the thesis that mathematics is reducible to logic, and hence nothing but a part of logic.[7]:41 Logicists hold that mathematics can be known *a priori*, but suggest that our knowledge of mathematics is just part of our knowledge of logic in general, and is thus analytic, not requiring any special faculty of mathematical intuition. In this view, logic is the proper foundation of mathematics, and all mathematical statements are necessary logical truths.

Rudolf Carnap (1931) presents the logicist thesis in two parts:[7]

1. The *concepts* of mathematics can be derived from logical concepts through explicit definitions.

2. The *theorems* of mathematics can be derived from logical axioms through purely logical deduction.

Gottlob Frege was the founder of logicism. In his seminal *Die Grundgesetze der Arithmetik* (*Basic Laws of Arithmetic*) he built up arithmetic from a system of logic with a general principle of comprehension, which he called "Basic Law V" (for concepts F and G, the extension of F equals the extension of G if and only if for all objects a, Fa if and only if Ga), a principle that he took to be acceptable as part of logic.

Frege's construction was flawed. Russell discovered that Basic Law V is inconsistent (this is Russell's paradox). Frege abandoned his logicist program soon after this, but it was continued by Russell and Whitehead. They attributed the paradox to "vicious circularity" and built up what they called ramified type theory to deal with it. In this system, they were eventually able to build up much of modern mathematics but in an altered, and excessively complex form (for example, there were different natural numbers in each type, and there were infinitely many types). They also had to make several compromises in order to develop so much of mathematics, such as an "axiom of reducibility". Even Russell said that this axiom did not really belong to logic.

Modern logicists (like Bob Hale, Crispin Wright, and perhaps others) have returned to a program closer to Frege's. They have abandoned Basic Law V in favor of abstraction principles such as Hume's principle (the number of objects falling under the concept F equals the number of objects falling under the concept G if and only if the extension of F and the extension of G can be put into one-to-one correspondence). Frege required Basic Law V to be able to give an explicit definition of the numbers, but all the properties of numbers can be derived from Hume's principle. This would not have been enough for Frege because (to paraphrase him) it does not exclude the possibility that the number 3 is in fact Julius Caesar. In addition, many of the weakened principles that they have had to adopt to replace Basic Law V no longer seem so obviously analytic, and thus purely logical.

2.4.5   Formalism

Main article: Formalism (mathematics)

Formalism holds that mathematical statements may be thought of as statements about the consequences of certain string manipulation rules. For example, in the "game" of Euclidean geometry (which is seen as consisting of some strings called "axioms", and some "rules of inference" to generate new strings from given ones), one can prove that the Pythagorean theorem holds (that is, you can generate the string corresponding to the Pythagorean theorem). According to formalism, mathematical truths are not about numbers and sets and triangles and the like—in fact, they aren't "about" anything at all.

Another version of formalism is often known as deductivism. In deductivism, the Pythagorean theorem is not an absolute truth, but a relative one: *if* you assign meaning to the strings in such a way that the rules of the game become true (i.e., true statements are assigned to the axioms and the rules of inference are truth-preserving), *then* you have to accept the theorem, or, rather, the interpretation you have given it must be a true statement. The same is held to be true for all other mathematical statements. Thus, formalism need not mean that mathematics is nothing more than a meaningless symbolic game. It is usually hoped that there exists some interpretation in which the rules of the game hold. (Compare this position to structuralism.) But it does allow the working mathematician to continue in his or her work and leave such problems to the philosopher or scientist. Many formalists would say that in practice, the axiom systems to be studied will be suggested by the demands of science or other areas of mathematics.

A major early proponent of formalism was David Hilbert, whose program was intended to be a complete and consistent axiomatization of all of mathematics. Hilbert aimed to show the consistency of mathematical systems from the assumption that the "finitary arithmetic" (a subsystem of the usual arithmetic of the positive integers, chosen to be philosophically uncontroversial) was consistent. Hilbert's goals of creating a system of mathematics that is both complete and consistent were dealt a fatal blow by the second of Gödel's incompleteness theorems, which states that sufficiently expressive consistent axiom systems can never prove their own consistency. Since any such axiom system would contain the finitary arithmetic as a subsystem, Gödel's theorem implied that it would be impossible to prove the system's consistency relative to that (since it would then prove its own consistency, which Gödel had shown was impossible). Thus, in order to show that any axiomatic system of mathematics is in fact consistent, one needs to first assume the consistency of a system of mathematics that is in a sense stronger than the system to be proven consistent.

Hilbert was initially a deductivist, but, as may be clear from above, he considered certain metamathematical methods to yield intrinsically meaningful results and was a realist with respect to the finitary arithmetic. Later, he held the opinion that there was no other meaningful mathematics whatsoever, regardless of interpretation.

Other formalists, such as Rudolf Carnap, Alfred Tarski, and Haskell Curry, considered mathematics to be the investigation

David Hilbert

of formal axiom systems. Mathematical logicians study formal systems but are just as often realists as they are formalists.

Formalists are relatively tolerant and inviting to new approaches to logic, non-standard number systems, new set theories etc. The more games we study, the better. However, in all three of these examples, motivation is drawn from existing mathematical or philosophical concerns. The "games" are usually not arbitrary.

The main critique of formalism is that the actual mathematical ideas that occupy mathematicians are far removed from the string manipulation games mentioned above. Formalism is thus silent on the question of which axiom systems ought to be studied, as none is more meaningful than another from a formalistic point of view.

Recently, some formalist mathematicians have proposed that all of our *formal* mathematical knowledge should be systematically encoded in computer-readable formats, so as to facilitate automated proof checking of mathematical proofs and the use of interactive theorem proving in the development of mathematical theories and computer software. Because of their close connection with computer science, this idea is also advocated by mathematical intuitionists and constructivists in the "computability" tradition (see below). See QED project for a general overview.

2.4.6   Conventionalism

The French mathematician Henri Poincaré was among the first to articulate a conventionalist view. Poincaré's use of non-Euclidean geometries in his work on differential equations convinced him that Euclidean geometry should not be regarded as *a priori* truth. He held that axioms in geometry should be chosen for the results they produce, not for their apparent coherence with human intuitions about the physical world.

2.4.7   Psychologism

Psychologism in the philosophy of mathematics is the position that mathematical concepts and/or truths are grounded in, derived from or explained by psychological facts (or laws).

John Stuart Mill seems to have been an advocate of a type of logical psychologism, as were many 19th-century German logicians such as Sigwart and Erdmann as well as a number of psychologists, past and present: for example, Gustave Le Bon. Psychologism was famously criticized by Frege in his *The Foundations of Arithmetic*, and many of his works and essays, including his review of Husserl's *Philosophy of Arithmetic*. Edmund Husserl, in the first volume of his *Logical Investigations*, called "The Prolegomena of Pure Logic", criticized psychologism thoroughly and sought to distance himself from it. The "Prolegomena" is considered a more concise, fair, and thorough refutation of psychologism than the criticisms made by Frege, and also it is considered today by many as being a memorable refutation for its decisive blow to psychologism. Psychologism was also criticized by Charles Sanders Peirce and Maurice Merleau-Ponty.

2.4.8   Intuitionism

Main article: Mathematical intuitionism

In mathematics, intuitionism is a program of methodological reform whose motto is that "there are no non-experienced mathematical truths" (L.E.J. Brouwer). From this springboard, intuitionists seek to reconstruct what they consider to be the corrigible portion of mathematics in accordance with Kantian concepts of being, becoming, intuition, and knowledge. Brouwer, the founder of the movement, held that mathematical objects arise from the *a priori* forms of the volitions that inform the perception of empirical objects.[8]

A major force behind intuitionism was L.E.J. Brouwer, who rejected the usefulness of formalized logic of any sort for mathematics. His student Arend Heyting postulated an intuitionistic logic, different from the classical Aristotelian logic; this logic does not contain the law of the excluded middle and therefore frowns upon proofs by contradiction. The axiom of choice is also rejected in most intuitionistic set theories, though in some versions it is accepted. Important work was later done by Errett Bishop, who managed to prove versions of the most important theorems in real analysis within this framework.

In intuitionism, the term "explicit construction" is not cleanly defined, and that has led to criticisms. Attempts have been made to use the concepts of Turing machine or computable function to fill this gap, leading to the claim that only questions regarding the behavior of finite algorithms are meaningful and should be investigated in mathematics. This has led to the study of the computable numbers, first introduced by Alan Turing. Not surprisingly, then, this approach to mathematics is sometimes associated with theoretical computer science.

Constructivism

Main article: Mathematical constructivism

Like intuitionism, constructivism involves the regulative principle that only mathematical entities which can be explicitly constructed in a certain sense should be admitted to mathematical discourse. In this view, mathematics is an exercise of the human intuition, not a game played with meaningless symbols. Instead, it is about entities that we can create directly through mental activity. In addition, some adherents of these schools reject non-constructive proofs, such as a proof by contradiction.

Finitism

Finitism is an extreme form of constructivism, according to which a mathematical object does not exist unless it can be constructed from natural numbers in a finite number of steps. In her book *Philosophy of Set Theory*, Mary Tiles characterized those who allow countably infinite objects as classical finitists, and those who deny even countably infinite objects as strict finitists.

The most famous proponent of finitism was Leopold Kronecker,[9] who said:

> God created the natural numbers, all else is the work of man.

Ultrafinitism is an even more extreme version of finitism, which rejects not only infinities but finite quantities that cannot feasibly be constructed with available resources.

2.4.9 Structuralism

Main article: Mathematical structuralism

Structuralism is a position holding that mathematical theories describe structures, and that mathematical objects are exhaustively defined by their *places* in such structures, consequently having no intrinsic properties. For instance, it would maintain that all that needs to be known about the number 1 is that it is the first whole number after 0. Likewise all the other whole numbers are defined by their places in a structure, the number line. Other examples of mathematical objects might include lines and planes in geometry, or elements and operations in abstract algebra.

Structuralism is an epistemologically realistic view in that it holds that mathematical statements have an objective truth value. However, its central claim only relates to what *kind* of entity a mathematical object is, not to what kind of *existence* mathematical objects or structures have (not, in other words, to their ontology). The kind of existence mathematical objects have would clearly be dependent on that of the structures in which they are embedded; different sub-varieties of structuralism make different ontological claims in this regard.[10]

The *Ante Rem*, or fully realist, variation of structuralism has a similar ontology to Platonism in that structures are held to have a real but abstract and immaterial existence. As such, it faces the usual problems of explaining the interaction between such abstract structures and flesh-and-blood mathematicians.

In Re, or moderately realistic, structuralism is the equivalent of Aristotelian realism. Structures are held to exist inasmuch as some concrete system exemplifies them. This incurs the usual issues that some perfectly legitimate structures might accidentally happen not to exist, and that a finite physical world might not be "big" enough to accommodate some otherwise legitimate structures.

The *Post Res* or eliminative variant of structuralism is anti-realist about structures in a way that parallels nominalism. According to this view mathematical *systems* exist, and have structural features in common. If something is true of a structure, it will be true of all systems exemplifying the structure. However, it is merely convenient to talk of structures being "held in common" between systems: they in fact have no independent existence.

2.4.10 Embodied mind theories

Embodied mind theories hold that mathematical thought is a natural outgrowth of the human cognitive apparatus which finds itself in our physical universe. For example, the abstract concept of number springs from the experience of counting discrete objects. It is held that mathematics is not universal and does not exist in any real sense, other than in human brains. Humans construct, but do not discover, mathematics.

With this view, the physical universe can thus be seen as the ultimate foundation of mathematics: it guided the evolution of the brain and later determined which questions this brain would find worthy of investigation. However, the human mind has no special claim on reality or approaches to it built out of math. If such constructs as Euler's identity are true then they are true as a map of the human mind and cognition.

Embodied mind theorists thus explain the effectiveness of mathematics—mathematics was constructed by the brain in order to be effective in this universe.

The most accessible, famous, and infamous treatment of this perspective is *Where Mathematics Comes From*, by George Lakoff and Rafael E. Núñez. In addition, mathematician Keith Devlin has investigated similar concepts with his book *The Math Instinct*, as has neuroscientist Stanislas Dehaene with his book *The Number Sense*. For more on the philosophical ideas that inspired this perspective, see cognitive science of mathematics.

New empiricism

A more recent empiricism returns to the principle of the English empiricists of the 18th and 19th centuries, in particular John Stuart Mill, who asserted that all knowledge comes to us from observation through the senses. This applies not only to matters of fact, but also to "relations of ideas", as Hume called them: the structures of logic which interpret, organize and abstract observations.

To this principle it adds a materialist connection: all the processes of logic which interpret, organize and abstract observations, are physical phenomena which take place in real time and physical space: namely, in the brains of human beings. Abstract objects, such as mathematical objects, are ideas, which in turn exist as electrical and chemical states of the billions of neurons in the human brain.

This second concept is reminiscent of the social constructivist approach, which holds that mathematics is produced by humans rather than being "discovered" from abstract, *a priori* truths. However, it differs sharply from the constructivist implication that humans arbitrarily construct mathematical principles that have no inherent truth but which instead are created on a conveniency basis. On the contrary, new empiricism shows how mathematics, although constructed by humans, follows rules and principles that will be agreed on by all who participate in the process, with the result that everyone practicing mathematics comes up with the same answer—except in those areas where there is philosophical disagreement on the meaning of fundamental concepts. This is because the new empiricism perceives this agreement as being a physical phenomenon, one which is observed by other humans in the same way that other physical phenomena, like the motions of inanimate bodies, or the chemical interaction of various elements, are observed.

Combining the materialist principle with Millisian epistemology evades the principal difficulty with classical empiricism— that all knowledge comes from the senses. That difficulty lies in the observation that mathematical truths based on logical deduction appear to be more certainly true than knowledge of the physical world itself. (The physical world in this case is taken to mean the portion of it lying outside the human brain.)

Kant argued that the structures of logic which organize, interpret and abstract observations were built into the human mind and were true and valid *a priori*. Mill, on the contrary, said that we believe them to be true because we have enough individual instances of their truth to generalize: in his words, "From instances we have observed, we feel warranted in concluding that what we found true in those instances holds in all similar ones, past, present and future, however numerous they may be".[11] Although the psychological or epistemological specifics given by Mill through which we build our logical apparatus may not be completely warranted, his explanation still nonetheless manages to demonstrate that there is no way around Kant's *a priori* logic. To recant Mill's original idea in an empiricist twist: "*Indeed, the very principles of logical deduction are true because we observe that using them leads to true conclusions*", which is itself an *a priori* presupposition.

If all this is true, then where do the world senses come in? The early empiricists all stumbled over this point. Hume asserted that all knowledge comes from the senses, and then gave away the ballgame by excepting abstract propositions, which he called "relations of ideas". These, he said, were absolutely true (although the mathematicians who thought them up, being human, might get them wrong). Mill, on the other hand, tried to deny that abstract ideas exist outside the physical world: all numbers, he said, "must be numbers of something: there are no such things as numbers in the abstract". When we count to eight or add five and three we are really counting spoons or bumblebees. "All things possess quantity", he said, so that propositions concerning numbers are propositions concerning "all things whatever". But then in almost a contradiction of himself he went on to acknowledge that numerical and algebraic expressions are not necessarily attached to real world objects: they "do not excite in our minds ideas of any things in particular". Mill's low reputation as a philosopher of logic, and the low estate of empiricism in the century and a half following him, derives from this failed attempt to link abstract thoughts to the physical world, when it may be more plausibly arguable that abstraction consists precisely of separating the thought from its physical foundations.

The conundrum created by our certainty that abstract deductive propositions, if valid (i.e. if we can "prove" them), are true, exclusive of observation and testing in the physical world, gives rise to a further reflection ... What if thoughts themselves, and the minds that create them, are physical objects, existing only in the physical world?

This would reconcile the contradiction between our belief in the certainty of abstract deductions and the empiricist principle that knowledge comes from observation of individual instances. We know that Euler's equation is true because every time a human mind derives the equation, it gets the same result, unless it has made a mistake, which can be acknowledged and corrected. We observe this phenomenon, and we extrapolate to the general proposition that it is always true.

This applies not only to physical principles, like the law of gravity, but to abstract phenomena that we observe only in human brains: in ours and in those of others.

Aristotelian realism

Main article: Aristotle's theory of universals

Similar to empiricism in emphasizing the relation of mathematics to the real world, Aristotelian realism holds that mathematics studies properties such as symmetry, continuity and order that can be literally realized in the physical world (or in any other world there might be). It contrasts with Platonism in holding that the objects of mathematics, such as numbers, do not exist in an "abstract" world but can be physically realized. For example, the number 4 is realized in the relation between a heap of parrots and the universal "being a parrot" that divides the heap into so many parrots.[12] Aristotelian realism is defended by James Franklin and the Sydney School in the philosophy of mathematics and is close to the view of Penelope Maddy that when an egg carton is opened, a set of three eggs is perceived (that is, a mathematical entity realized in the physical world).[13] A problem for Aristotelian realism is what account to give of higher infinities, which may not be realizable in the physical world.

2.4.11 Fictionalism

Fictionalism in mathematics was brought to fame in 1980 when Hartry Field published *Science Without Numbers*, which rejected and in fact reversed Quine's indispensability argument. Where Quine suggested that mathematics was indispensable for our best scientific theories, and therefore should be accepted as a body of truths talking about independently existing entities, Field suggested that mathematics was dispensable, and therefore should be considered as a body of falsehoods not talking about anything real. He did this by giving a complete axiomatization of Newtonian mechanics that didn't reference numbers or functions at all. He started with the "betweenness" of Hilbert's axioms to characterize space without coordinatizing it, and then added extra relations between points to do the work formerly done by vector fields. Hilbert's geometry is mathematical, because it talks about abstract points, but in Field's theory, these points are the concrete points of physical space, so no special mathematical objects at all are needed.

Having shown how to do science without using numbers, Field proceeded to rehabilitate mathematics as a kind of useful fiction. He showed that mathematical physics is a conservative extension of his non-mathematical physics (that is, every physical fact provable in mathematical physics is already provable from Field's system), so that mathematics is a reliable process whose physical applications are all true, even though its own statements are false. Thus, when doing mathematics, we can see ourselves as telling a sort of story, talking as if numbers existed. For Field, a statement like "2 + 2 = 4" is just as fictitious as "Sherlock Holmes lived at 221B Baker Street"—but both are true according to the relevant fictions.

By this account, there are no metaphysical or epistemological problems special to mathematics. The only worries left are the general worries about non-mathematical physics, and about fiction in general. Field's approach has been very influential, but is widely rejected. This is in part because of the requirement of strong fragments of second-order logic to carry out his reduction, and because the statement of conservativity seems to require quantification over abstract models or deductions.

2.4.12   Social constructivism or social realism

Social constructivism or *social realism* theories see mathematics primarily as a social construct, as a product of culture, subject to correction and change. Like the other sciences, mathematics is viewed as an empirical endeavor whose results are constantly evaluated and may be discarded. However, while on an empiricist view the evaluation is some sort of comparison with "reality", social constructivists emphasize that the direction of mathematical research is dictated by the fashions of the social group performing it or by the needs of the society financing it. However, although such external forces may change the direction of some mathematical research, there are strong internal constraints—the mathematical traditions, methods, problems, meanings and values into which mathematicians are enculturated—that work to conserve the historically defined discipline.

This runs counter to the traditional beliefs of working mathematicians, that mathematics is somehow pure or objective. But social constructivists argue that mathematics is in fact grounded by much uncertainty: as mathematical practice evolves, the status of previous mathematics is cast into doubt, and is corrected to the degree it is required or desired by the current mathematical community. This can be seen in the development of analysis from reexamination of the calculus of Leibniz and Newton. They argue further that finished mathematics is often accorded too much status, and folk mathematics not enough, due to an overemphasis on axiomatic proof and peer review as practices. However, this might be seen as merely saying that rigorously proven results are overemphasized, and then "look how chaotic and uncertain the rest of it all is!"

The social nature of mathematics is highlighted in its subcultures. Major discoveries can be made in one branch of mathematics and be relevant to another, yet the relationship goes undiscovered for lack of social contact between mathematicians. Social constructivists argue each speciality forms its own epistemic community and often has great difficulty communicating, or motivating the investigation of unifying conjectures that might relate different areas of mathematics. Social constructivists see the process of "doing mathematics" as actually creating the meaning, while social realists see a deficiency either of human capacity to abstractify, or of human's cognitive bias, or of mathematicians' collective intelligence as preventing the comprehension of a real universe of mathematical objects. Social constructivists sometimes reject the search for foundations of mathematics as bound to fail, as pointless or even meaningless.

Contributions to this school have been made by Imre Lakatos and Thomas Tymoczko, although it is not clear that either would endorse the title. More recently Paul Ernest has explicitly formulated a social constructivist philosophy of mathematics.[14] Some consider the work of Paul Erdős as a whole to have advanced this view (although he personally rejected it) because of his uniquely broad collaborations, which prompted others to see and study "mathematics as a social activity", e.g., via the Erdős number. Reuben Hersh has also promoted the social view of mathematics, calling it a "humanistic" approach,[15] similar to but not quite the same as that associated with Alvin White;[16] one of Hersh's co-authors, Philip J. Davis, has expressed sympathy for the social view as well.

A criticism of this approach is that it is trivial, based on the trivial observation that mathematics is a human activity. To observe that rigorous proof comes only after unrigorous conjecture, experimentation and speculation is true, but it is trivial and no-one would deny this. So it's a bit of a stretch to characterize a philosophy of mathematics in this way, on something trivially true. The calculus of Leibniz and Newton was reexamined by mathematicians such as Weierstrass in order to rigorously prove the theorems thereof. There is nothing special or interesting about this, as it fits in with the more general trend of unrigorous ideas which are later made rigorous. There needs to be a clear distinction between the objects of study of mathematics and the study of the objects of study of mathematics. The former doesn't seem to change a great deal; the latter is forever in flux. The latter is what the social theory is about, and the former is what Platonism *et al.* are about.

However, this criticism is rejected by supporters of the social constructivist perspective because it misses the point that the very objects of mathematics are social constructs. These objects, it asserts, are primarily semiotic objects existing in the sphere of human culture, sustained by social practices (after Wittgenstein) that utilize physically embodied signs and give rise to intrapersonal (mental) constructs. Social constructivists view the reification of the sphere of human culture into a Platonic realm, or some other heaven-like domain of existence beyond the physical world, a long-standing category error.

2.4.13 Beyond the traditional schools

Rather than focus on narrow debates about the true nature of mathematical truth, or even on practices unique to mathematicians such as the proof, a growing movement from the 1960s to the 1990s began to question the idea of seeking foundations or finding any one right answer to why mathematics works. The starting point for this was Eugene Wigner's famous 1960 paper *The Unreasonable Effectiveness of Mathematics in the Natural Sciences*, in which he argued that the happy coincidence of mathematics and physics being so well matched seemed to be unreasonable and hard to explain.

The embodied-mind or cognitive school and the social school were responses to this challenge, but the debates raised were difficult to confine to those.

Quasi-empiricism

One parallel concern that does not actually challenge the schools directly but instead questions their focus is the notion of quasi-empiricism in mathematics. This grew from the increasingly popular assertion in the late 20th century that no one foundation of mathematics could be ever proven to exist. It is also sometimes called "postmodernism in mathematics" although that term is considered overloaded by some and insulting by others. Quasi-empiricism argues that in doing their research, mathematicians test hypotheses as well as prove theorems. A mathematical argument can transmit falsity from the conclusion to the premises just as well as it can transmit truth from the premises to the conclusion. Quasi-empiricism was developed by Imre Lakatos, inspired by the philosophy of science of Karl Popper.

Lakatos' philosophy of mathematics is sometimes regarded as a kind of social constructivism, but this was not his intention.

Such methods have always been part of folk mathematics by which great feats of calculation and measurement are sometimes achieved. Indeed, such methods may be the only notion of proof a culture has.

Hilary Putnam has argued that any theory of mathematical realism would include quasi-empirical methods. He proposed that an alien species doing mathematics might well rely on quasi-empirical methods primarily, being willing often to forgo rigorous and axiomatic proofs, and still be doing mathematics—at perhaps a somewhat greater risk of failure of their calculations. He gave a detailed argument for this in *New Directions*.[17]

Popper's "two senses" theory

Realist and constructivist theories are normally taken to be contraries. However, Karl Popper[18] argued that a number statement such as "2 apples + 2 apples = 4 apples" can be taken in two senses. In one sense it is irrefutable and logically true. In the second sense it is factually true and falsifiable. Another way of putting this is to say that a single number statement can express two propositions: one of which can be explained on constructivist lines; the other on realist lines.[19]

Language

Main article: Philosophy of language

Innovations in the philosophy of language during the 20th century renewed interest in whether mathematics is, as is often said, the *language* of science. Although some mathematicians and philosophers would accept the statement "mathematics is a language", linguists believe that the implications of such a statement must be considered. For example, the tools of linguistics are not generally applied to the symbol systems of mathematics, that is, mathematics is studied in a markedly different way than other languages. If mathematics is a language, it is a different type of language than natural languages. Indeed, because of the need for clarity and specificity, the language of mathematics is far more constrained than natural languages studied by linguists. However, the methods developed by Frege and Tarski for the study of mathematical language have been extended greatly by Tarski's student Richard Montague and other linguists working in formal semantics to show that the distinction between mathematical language and natural language may not be as great as it seems.

2.5   Arguments

2.5.1   Indispensability argument for realism

This argument, associated with Willard Quine and Hilary Putnam, is considered by Stephen Yablo to be one of the most challenging arguments in favor of the acceptance of the existence of abstract mathematical entities, such as numbers and sets.[20] The form of the argument is as follows.

1. One must have ontological commitments to *all* entities that are indispensable to the best scientific theories, and to those entities *only* (commonly referred to as "all and only").

2. Mathematical entities are indispensable to the best scientific theories. Therefore,

3. One must have ontological commitments to mathematical entities.[21]

The justification for the first premise is the most controversial. Both Putnam and Quine invoke naturalism to justify the exclusion of all non-scientific entities, and hence to defend the "only" part of "all and only". The assertion that "all" entities postulated in scientific theories, including numbers, should be accepted as real is justified by confirmation holism. Since theories are not confirmed in a piecemeal fashion, but as a whole, there is no justification for excluding any of the entities referred to in well-confirmed theories. This puts the nominalist who wishes to exclude the existence of sets and non-Euclidean geometry, but to include the existence of quarks and other undetectable entities of physics, for example, in a difficult position.[21]

2.5.2   Epistemic argument against realism

The anti-realist "epistemic argument" against Platonism has been made by Paul Benacerraf and Hartry Field. Platonism posits that mathematical objects are *abstract* entities. By general agreement, abstract entities cannot interact causally with concrete, physical entities. ("the truth-values of our mathematical assertions depend on facts involving Platonic entities that reside in a realm outside of space-time"[22]) Whilst our knowledge of concrete, physical objects is based on our ability to perceive them, and therefore to causally interact with them, there is no parallel account of how mathematicians come to have knowledge of abstract objects.[23][24][25] ("An account of mathematical truth ... must be consistent with the possibility of mathematical knowledge."[26]) Another way of making the point is that if the Platonic world were to disappear, it would make no difference to the ability of mathematicians to generate proofs, etc., which is already fully accountable in terms of physical processes in their brains.

Field developed his views into fictionalism. Benacerraf also developed the philosophy of mathematical structuralism, according to which there are no mathematical objects. Nonetheless, some versions of structuralism are compatible with some versions of realism.

The argument hinges on the idea that a satisfactory naturalistic account of thought processes in terms of brain processes can be given for mathematical reasoning along with everything else. One line of defense is to maintain that this is false, so that mathematical reasoning uses some special intuition that involves contact with the Platonic realm. A modern form of this argument is given by Sir Roger Penrose.[27]

Another line of defense is to maintain that abstract objects are relevant to mathematical reasoning in a way that is non-causal, and not analogous to perception. This argument is developed by Jerrold Katz in his book *Realistic Rationalism*.

A more radical defense is denial of physical reality, i.e. the mathematical universe hypothesis. In that case, a mathematician's knowledge of mathematics is one mathematical object making contact with another.

2.6 Aesthetics

Many practicing mathematicians have been drawn to their subject because of a sense of beauty they perceive in it. One sometimes hears the sentiment that mathematicians would like to leave philosophy to the philosophers and get back to mathematics—where, presumably, the beauty lies.

In his work on the divine proportion, H.E. Huntley relates the feeling of reading and understanding someone else's proof of a theorem of mathematics to that of a viewer of a masterpiece of art—the reader of a proof has a similar sense of exhilaration at understanding as the original author of the proof, much as, he argues, the viewer of a masterpiece has a sense of exhilaration similar to the original painter or sculptor. Indeed, one can study mathematical and scientific writings as literature.

Philip J. Davis and Reuben Hersh have commented that the sense of mathematical beauty is universal amongst practicing mathematicians. By way of example, they provide two proofs of the irrationality of the $\sqrt{2}$. The first is the traditional proof by contradiction, ascribed to Euclid; the second is a more direct proof involving the fundamental theorem of arithmetic that, they argue, gets to the heart of the issue. Davis and Hersh argue that mathematicians find the second proof more aesthetically appealing because it gets closer to the nature of the problem.

Paul Erdős was well known for his notion of a hypothetical "Book" containing the most elegant or beautiful mathematical proofs. There is not universal agreement that a result has one "most elegant" proof; Gregory Chaitin has argued against this idea.

Philosophers have sometimes criticized mathematicians' sense of beauty or elegance as being, at best, vaguely stated. By the same token, however, philosophers of mathematics have sought to characterize what makes one proof more desirable than another when both are logically sound.

Another aspect of aesthetics concerning mathematics is mathematicians' views towards the possible uses of mathematics for purposes deemed unethical or inappropriate. The best-known exposition of this view occurs in G.H. Hardy's book *A Mathematician's Apology*, in which Hardy argues that pure mathematics is superior in beauty to applied mathematics precisely because it cannot be used for war and similar ends. Some later mathematicians have characterized Hardy's views as mildly dated, with the applicability of number theory to modern-day cryptography.

2.7 See also

2.7.1 Related works

2.7.2 Historical topics

- History and philosophy of science

- History of mathematics

- History of philosophy

2.8 Notes

[1] Maziars, Edward A. (1969). "Problems in the Philosophy of Mathematics (Book Review)". *Philosophy of Science* **36** (3): 325.. For example, when Edward Maziars proposes in a 1969 book review *"to distinguish philosophical mathematics (which is primarily a specialised task for a mathematician) from mathematical philosophy (which ordinarily may be the philosopher's metier)"*, he uses the term *mathematical philosophy* as being synonymous with *philosophy of mathematics*.

[2] Kleene, Stephen (1971). *Introduction to Metamathematics*. Amsterdam, Netherlands: North-Holland Publishing Company. p. 5.

[3] Mac Lane, Saunders (1998), *Categories for the Working Mathematician*, 2nd edition, Springer-Verlag, New York, NY.

[4]   • Putnam, Hilary (1967), "Mathematics Without Foundations", *Journal of Philosophy* 64/1, 5-22. Reprinted, pp. 168–184 in W.D. Hart (ed., 1996).

[5] Tegmark, Max (February 2008). "The Mathematical Universe". *Foundations of Physics* **38** (2): 101–150. arXiv:0704.0646. Bibcode:2008FoPh...38..101T. doi:10.1007/s10701-007-9186-9.

[6] Tegmark (1998), p. 1.

[7] Carnap, Rudolf (1931), "Die logizistische Grundlegung der Mathematik", *Erkenntnis* 2, 91-121. Republished, "The Logicist Foundations of Mathematics", E. Putnam and G.J. Massey (trans.), in Benacerraf and Putnam (1964). Reprinted, pp. 41–52 in Benacerraf and Putnam (1983).

[8] Audi, Robert (1999), *The Cambridge Dictionary of Philosophy*, Cambridge University Press, Cambridge, UK, 1995. 2nd edition. Page 542.

[9] From an 1886 lecture at the 'Berliner Naturforscher-Versammlung', according to H. M. Weber's memorial article, as quoted and translated in Gonzalez Cabillon, Julio (2000-02-03). "FOM: What were Kronecker's f.o.m.?". Retrieved 2008-07-19. Gonzalez gives as the sources for the memorial article, the following: 'Weber, H: "Leopold Kronecker", _Jahresberichte der Deutschen Mathematiker Vereinigung_, vol ii (1893) pp 5-31. Cf page 19. See also _Mathematische Annalen_ vol xliii (1893) pp 1-25'.

[10] Brown, James (2008). *Philosophy of Mathematics*. New York: Routledge. ISBN 978-0-415-96047-2.

[11] A System of Logic Ratiocinative and Inductive, The Collected Works of John Stuart Mill published by the University of Toronto Press in 1973. Book II, Chapter vi, Section 2 (Toronto edition 1975, Vol.7, p. 254)

[12] Franklin, James (2014), "An Aristotelian Realist Philosophy of Mathematics", Palgrave Macmillan, Basingstoke; Franklin, James (2011), "Aristotelianism in the philosophy of mathematics," *Studia Neoaristotelica* 8, 3-15.

[13] Maddy, Penelope (1990), *Realism in Mathematics*, Oxford University Press, Oxford, UK.

[14] Ernest, Paul. "Is Mathematics Discovered or Invented?". University of Exeter. Retrieved 2008-12-26.

[15] Hersh, Reuben (February 10, 1997). *What Kind of a Thing is a Number?*. Interview with John Brockman. Edge Foundation. Retrieved 2008-12-26.

[16] "Humanism and Mathematics Education". *Math Forum*. Humanistic Mathematics Network Journal. Retrieved 2008-12-26.

[17] Tymoczko, Thomas (1998), *New Directions in the Philosophy of Mathematics*. ISBN 978-0691034980.

[18] Popper, Karl Raimund (1946) Aristotelian Society Supplementary Volume XX.

[19] Gregory, Frank Hutson (1996) Arithmetic and Reality: A Development of Popper's Ideas. City University of Hong Kong. Republished in Philosophy of Mathematics Education Journal No. 26 (December 2011)

[20] Yablo, S. (November 8, 1998). "A Paradox of Existence".

[21] Putnam, H. *Mathematics, Matter and Method. Philosophical Papers, vol. 1*. Cambridge: Cambridge University Press, 1975. 2nd. ed., 1985.

[22] Field, Hartry, 1989, Realism, Mathematics, and Modality, Oxford: Blackwell, p. 68

[23] "Since abstract objects are outside the nexus of causes and effects, and thus perceptually inaccessible, they cannot be known through their effects on us" Katz, J. *Realistic Rationalism*, p15

[24] ,Philosophy Now: *Mathematical_Knowledge_A_Dilemma Mathematical Knowledge: A dilemma*

[25] Standard Encyclopaedia of Philosophy

[26] Benacceraf, 1973, p409

[27] Review of The Emperor's New Mind

2.9 Further reading

- Aristotle, "Prior Analytics", Hugh Tredennick (trans.), pp. 181–531 in *Aristotle, Volume 1*, Loeb Classical Library, William Heinemann, London, UK, 1938.

- Benacerraf, Paul, and Putnam, Hilary (eds., 1983), *Philosophy of Mathematics, Selected Readings*, 1st edition, Prentice-Hall, Englewood Cliffs, NJ, 1964. 2nd edition, Cambridge University Press, Cambridge, UK, 1983.

- Berkeley, George (1734), *The Analyst; or, a Discourse Addressed to an Infidel Mathematician. Wherein It is examined whether the Object, Principles, and Inferences of the modern Analysis are more distinctly conceived, or more evidently deduced, than Religious Mysteries and Points of Faith*, London & Dublin. Online text, David R. Wilkins (ed.), Eprint.

- Bourbaki, N. (1994), *Elements of the History of Mathematics*, John Meldrum (trans.), Springer-Verlag, Berlin, Germany.

- Chandrasekhar, Subrahmanyan (1987), *Truth and Beauty. Aesthetics and Motivations in Science*, University of Chicago Press, Chicago, IL.

- Colyvan, Mark (2004), "Indispensability Arguments in the Philosophy of Mathematics", *Stanford Encyclopedia of Philosophy*, Edward N. Zalta (ed.), Eprint.

- Davis, Philip J. and Hersh, Reuben (1981), *The Mathematical Experience*, Mariner Books, New York, NY.

- Devlin, Keith (2005), *The Math Instinct: Why You're a Mathematical Genius (Along with Lobsters, Birds, Cats, and Dogs)*, Thunder's Mouth Press, New York, NY.

- Dummett, Michael (1991 a), *Frege, Philosophy of Mathematics*, Harvard University Press, Cambridge, MA.

- Dummett, Michael (1991 b), *Frege and Other Philosophers*, Oxford University Press, Oxford, UK.

- Dummett, Michael (1993), *Origins of Analytical Philosophy*, Harvard University Press, Cambridge, MA.

- Ernest, Paul (1998), *Social Constructivism as a Philosophy of Mathematics*, State University of New York Press, Albany, NY.

- George, Alexandre (ed., 1994), *Mathematics and Mind*, Oxford University Press, Oxford, UK.

- Hadamard, Jacques (1949), *The Psychology of Invention in the Mathematical Field*, 1st edition, Princeton University Press, Princeton, NJ. 2nd edition, 1949. Reprinted, Dover Publications, New York, NY, 1954.

- Hardy, G.H. (1940), *A Mathematician's Apology*, 1st published, 1940. Reprinted, C.P. Snow (foreword), 1967. Reprinted, Cambridge University Press, Cambridge, UK, 1992.

- Hart, W.D. (ed., 1996), *The Philosophy of Mathematics*, Oxford University Press, Oxford, UK.

- Hendricks, Vincent F. and Hannes Leitgeb (eds.). *Philosophy of Mathematics: 5 Questions*, New York: Automatic Press / VIP, 2006.

- Huntley, H.E. (1970), *The Divine Proportion: A Study in Mathematical Beauty*, Dover Publications, New York, NY.

- Irvine, A., ed (2009), *The Philosophy of Mathematics*, in *Handbook of the Philosophy of Science* series, North-Holland Elsevier, Amsterdam.

- Klein, Jacob (1968), *Greek Mathematical Thought and the Origin of Algebra*, Eva Brann (trans.), MIT Press, Cambridge, MA, 1968. Reprinted, Dover Publications, Mineola, NY, 1992.

- Kline, Morris (1959), *Mathematics and the Physical World*, Thomas Y. Crowell Company, New York, NY, 1959. Reprinted, Dover Publications, Mineola, NY, 1981.

- Kline, Morris (1972), *Mathematical Thought from Ancient to Modern Times*, Oxford University Press, New York, NY.

- König, Julius (Gyula) (1905), "Über die Grundlagen der Mengenlehre und das Kontinuumproblem", *Mathematische Annalen* 61, 156-160. Reprinted, "On the Foundations of Set Theory and the Continuum Problem", Stefan Bauer-Mengelberg (trans.), pp. 145–149 in Jean van Heijenoort (ed., 1967).

- Körner, Stephan, *The Philosophy of Mathematics, An Introduction*. Harper Books, 1960.

- Lakoff, George, and Núñez, Rafael E. (2000), *Where Mathematics Comes From: How the Embodied Mind Brings Mathematics into Being*, Basic Books, New York, NY.

- Lakatos, Imre 1976 *Proofs and Refutations:The Logic of Mathematical Discovery* (Eds) J. Worrall & E. Zahar Cambridge University Press

- Lakatos, Imre 1978 *Mathematics, Science and Epistemology: Philosophical Papers* Volume 2 (Eds) J.Worrall & G.Currie Cambridge University Press

- Lakatos, Imre 1968 *Problems in the Philosophy of Mathematics* North Holland

- Leibniz, G.W., *Logical Papers* (1666–1690), G.H.R. Parkinson (ed., trans.), Oxford University Press, London, UK, 1966.

- Maddy, Penelope (1997), *Naturalism in Mathematics*, Oxford University Press, Oxford, UK.

- Maziarz, Edward A., and Greenwood, Thomas (1995), *Greek Mathematical Philosophy*, Barnes and Noble Books.

- Mount, Matthew, *Classical Greek Mathematical Philosophy*, .

- Parsons, Charles (2014). *Philosophy of Mathematics in the Twentieth Century: Selected Essays*. Cambridge, MA: Harvard University Press. ISBN 978-0-674-72806-6.

- Peirce, Benjamin (1870), "Linear Associative Algebra", § 1. See *American Journal of Mathematics* 4 (1881).

- Peirce, C.S., *Collected Papers of Charles Sanders Peirce*, vols. 1-6, Charles Hartshorne and Paul Weiss (eds.), vols. 7-8, Arthur W. Burks (ed.), Harvard University Press, Cambridge, MA, 1931 – 1935, 1958. Cited as CP (volume).(paragraph).

- Peirce, C.S., various pieces on mathematics and logic, many readable online through links at the Charles Sanders Peirce bibliography, especially under Books authored or edited by Peirce, published in his lifetime and the two sections following it.

- Plato, "The Republic, Volume 1", Paul Shorey (trans.), pp. 1–535 in *Plato, Volume 5*, Loeb Classical Library, William Heinemann, London, UK, 1930.

- Plato, "The Republic, Volume 2", Paul Shorey (trans.), pp. 1–521 in *Plato, Volume 6*, Loeb Classical Library, William Heinemann, London, UK, 1935.

- Resnik, Michael D. *Frege and the Philosophy of Mathematics*, Cornell University, 1980.

- Resnik, Michael (1997), *Mathematics as a Science of Patterns*, Clarendon Press, Oxford, UK, ISBN 978-0-19-825014-2

- Robinson, Gilbert de B. (1959), *The Foundations of Geometry*, University of Toronto Press, Toronto, Canada, 1940, 1946, 1952, 4th edition 1959.

- Raymond, Eric S. (1993), "The Utility of Mathematics", Eprint.

- Smullyan, Raymond M. (1993), *Recursion Theory for Metamathematics*, Oxford University Press, Oxford, UK.

- Russell, Bertrand (1919), *Introduction to Mathematical Philosophy*, George Allen and Unwin, London, UK. Reprinted, John G. Slater (intro.), Routledge, London, UK, 1993.

- Shapiro, Stewart (2000), *Thinking About Mathematics: The Philosophy of Mathematics*, Oxford University Press, Oxford, UK

- Strohmeier, John, and Westbrook, Peter (1999), *Divine Harmony, The Life and Teachings of Pythagoras*, Berkeley Hills Books, Berkeley, CA.

- Styazhkin, N.I. (1969), *History of Mathematical Logic from Leibniz to Peano*, MIT Press, Cambridge, MA.

- Tait, William W. (1986), "Truth and Proof: The Platonism of Mathematics", *Synthese* 69 (1986), 341-370. Reprinted, pp. 142–167 in W.D. Hart (ed., 1996).

- Tarski, A. (1983), *Logic, Semantics, Metamathematics: Papers from 1923 to 1938*, J.H. Woodger (trans.), Oxford University Press, Oxford, UK, 1956. 2nd edition, John Corcoran (ed.), Hackett Publishing, Indianapolis, IN, 1983.

- Ulam, S.M. (1990), *Analogies Between Analogies: The Mathematical Reports of S.M. Ulam and His Los Alamos Collaborators*, A.R. Bednarek and Françoise Ulam (eds.), University of California Press, Berkeley, CA.

- van Heijenoort, Jean (ed. 1967), *From Frege To Gödel: A Source Book in Mathematical Logic, 1879-1931*, Harvard University Press, Cambridge, MA.

- Wigner, Eugene (1960), "The Unreasonable Effectiveness of Mathematics in the Natural Sciences", *Communications on Pure and Applied Mathematics* 13(1): 1-14. Eprint

- Wilder, Raymond L. *Mathematics as a Cultural System*, Pergamon, 1980.

2.10 External links

- Philosophy of mathematics at PhilPapers

- Philosophy of mathematics at the Indiana Philosophy Ontology Project

- Philosophy of Mathematics entry by Leon Horsten in the *Stanford Encyclopedia of Philosophy*

- Philosophy of mathematics entry in the *Internet Encyclopedia of Philosophy*

- The London Philosophy Study Guide offers many suggestions on what to read, depending on the student's familiarity with the subject:

  - Philosophy of Mathematics
  - Mathematical Logic
  - Set Theory & Further Logic

- R.B. Jones' philosophy of mathematics page

- Philosophy of mathematics at DMOZ

- The Philosophy of Real Mathematics Blog

- Kaina Stoicheia by C.S. Peirce.

2.10.1 Journals

- Philosophia Mathematica journal

- The Philosophy of Mathematics Education Journal homepage

Chapter 3

Language of mathematics

The **language of mathematics** is the system used by mathematicians to communicate mathematical ideas among themselves. This language consists of a substrate of some natural language (for example English) using technical terms and grammatical conventions that are peculiar to mathematical discourse (see Mathematical jargon), supplemented by a highly specialized symbolic notation for mathematical formulas.

Like natural languages in general, discourse using the language of mathematics can employ a scala of registers. Research articles in academic journals use a more formal tone than oral exchanges over a scribbled-upon napkin in the university cafeteria.

3.1 What is a language?

Here are some definitions of language:

- *a systematic means of communicating by the use of sounds or conventional symbols*

- *a system of words used in a particular discipline*

- *a system of abstract codes which represent antecedent events and concepts* [1]

- *the code we all use to express ourselves and communicate to others* Speech & Language Therapy Glossary of Terms]

- *a set (finite or infinite) of sentences, each finite in length and constructed out of a finite set of elements* Noam Chomsky.

These definitions describe language in terms of the following components:

- A vocabulary of symbols or words

- A grammar consisting of rules of how these symbols may be used

- A 'syntax' or propositional structure, which places the symbols in linear structures.

- A 'Discourse' or 'narrative,' consisting of strings of syntactic propositions [2]

- A community of people who use and understand these symbols

- A range of meanings that can be communicated with these symbols

Each of these components is also found in the language of mathematics.

3.2 The vocabulary of mathematics

Mathematical notation has assimilated symbols from many different alphabets and typefaces. It also includes symbols that are specific to mathematics, such as

$\forall\ \exists\ \nabla\ \wedge\ \infty.$

Mathematical notation is central to the power of modern mathematics. Though the algebra of Al-Khwārizmī did not use such symbols, it solved equations using many more rules than are used today with symbolic notation, and had great difficulty working with multiple variables (which using symbolic notation can simply be called x, y, z , etc.). Sometimes formulas cannot be understood without a written or spoken explanation, but often they are sufficient by themselves, and sometimes they are difficult to read aloud or information is lost in the translation to words, as when several parenthetical factors are involved or when a complex structure like a matrix is manipulated.

Like any other profession, mathematics also has its own brand of technical terminology. In some cases, a word in general usage has a different and specific meaning within mathematics—examples are group, ring, field, category, term, and factor. For more examples, see Category:Mathematical terminology.

In other cases, specialist terms have been created which do not exist outside of mathematics—examples are tensor, fractal, functor. Mathematical statements have their own moderately complex taxonomy, being divided into axioms, conjectures, theorems, lemmas and corollaries. And there are stock phrases in mathematics, used with specific meanings, such as "*if and only if*", "*necessary and sufficient*" and "*without loss of generality*". Such phrases are known as mathematical jargon.

The vocabulary of mathematics also has visual elements. Diagrams are used informally on blackboards, as well as more formally in published work. When used appropriately, diagrams display schematic information more easily. Diagrams also help visually and aid intuitive calculations. Sometimes, as in a visual proof, a diagram even serves as complete justification for a proposition. A system of diagram conventions may evolve into a mathematical notation – for example, the Penrose graphical notation for tensor products.

3.3 The grammar of mathematics

The grammar used for mathematical discourse is essentially the grammar of the natural language used as substrate, but with several mathematics-specific peculiarities.

Most notably, the mathematical notation used for formulas has its own grammar, not dependent on a specific natural language, but shared internationally by mathematicians regardless of their mother tongues. This includes the conventions that the formulas are written predominantly left to right, also when the writing system of the substrate language is right-to-left, and that the Latin alphabet is commonly used for simple variables and parameters. A formula such as

$\sin x + a \cos 2x \geq 0$

is understood by Chinese and Israeli mathematicians alike.

Such mathematical formulas can be a part of speech in a natural-language phrase, or even assume the role of a full-fledged sentence. For example, the formula above, an equation, can be considered a sentence or sentential phrase in which the greater than or equal to symbol has the role of a verb. In careful speech, this can be made clear by pronouncing "$\geq$" as "is greater than or equal to", but in an informal context mathematicians may shorten this to "greater or equal" and yet handle this grammatically like a verb. A good example is the book title *Why does E = mc² ?*;[3] here, the equals sign has the role of an infinitive.

Mathematical formulas can be *vocalized* (spoken aloud). The vocalization system for formulas has to be learned, and is dependent on the underlying natural language. For example, when using English, the expression "f (x)" is conventionally pronounced "eff of eks", where the insertion of the preposition "of" is not suggested by the notation per se. The expression "$\frac{dy}{dx}$ ", on the other hand, is vocalized like "dee-why-dee-eks", with complete omission of the fraction bar, in other contexts often pronounced "over". The book title *Why does E = mc² ?* is said aloud as *Why does ee equal em see-squared?*.

Characteristic for mathematical discourse – both formal and informal – is the use of the inclusive first person plural "we" to mean: "the audience (or reader) together with the speaker (or author)".

3.4 The language community of mathematics

Mathematics is used by mathematicians, who form a global community composed of speakers of many languages. It is also used by students of mathematics. As mathematics is a part of primary education in almost all countries, almost all educated people have some exposure to pure mathematics. There are very few cultural dependencies or barriers in modern mathematics. There are international mathematics competitions, such as the International Mathematical Olympiad, and international co-operation between professional mathematicians is commonplace.

3.5 The meanings of mathematics

Mathematics is used to communicate information about a wide range of different subjects. Here are three broad categories:

- **Mathematics describes the real world**: many areas of mathematics originated with attempts to describe and solve real world phenomena - from measuring farms (geometry) to falling apples (calculus) to gambling (probability). Mathematics is widely used in modern physics and engineering, and has been hugely successful in helping us to understand more about the universe around us from its largest scales (physical cosmology) to its smallest (quantum mechanics). Indeed, the very success of mathematics in this respect has been a source of puzzlement for some philosophers (see The Unreasonable Effectiveness of Mathematics in the Natural Sciences by Eugene Wigner).

- **Mathematics describes abstract structures**: on the other hand, there are areas of pure mathematics which deal with abstract structures, which have no known physical counterparts at all. However, it is difficult to give any categorical examples here, as even the most abstract structures can be co-opted as models in some branch of physics (see Calabi-Yau spaces and string theory).

- **Mathematics describes mathematics**: mathematics can be used reflexively to describe itself—this is an area of mathematics called metamathematics.

Mathematics can communicate a range of meanings that is as wide as (although different from) that of a natural language. As English mathematician R.L.E. Schwarzenberger says:

> *My own attitude, which I share with many of my colleagues, is simply that mathematics is a language. Like English, or Latin, or Chinese, there are certain concepts for which mathematics is particularly well suited: it would be as foolish to attempt to write a love poem in the language of mathematics as to prove the Fundamental Theorem of Algebra using the English language.*

3.6 Alternative views

Some definitions of language, such as early versions of Charles Hockett's "design features" definition, emphasize the spoken nature of language. Mathematics would not qualify as a language under these definitions, as it is primarily a written form of communication (to see why, try reading Maxwell's equations out loud). However, these definitions would also disqualify sign languages, which are now recognized as languages in their own right, independent of spoken language.

Other linguists believe no valid comparison can be made between mathematics and language, because they are simply too different:

> *Mathematics would appear to be both more and less than a language for while being limited in its linguistic capabilities it also seems to involve a form of thinking that has something in common with art and music.* -
> Ford & Peat (1988)

3.7 See also

- Formulario mathematico
- Linguistics
- Philosophy of language

3.8 References

[1] Syntax: An Introduction, Volume 1 Talmy GivónJohn Benjamins Publishing, 2001

[2] Syntax: An Introduction, Volume 1 Talmy Givón John Benjamins Publishing, 2001

[3] Brian Cox; Jeff Forshaw (2010). *Why does E = mc² ? (and why should we care?)*. Da Capo Press. ISBN 978-0-306-81876-9.

- Knight, Isabel F. (1968). *The Geometric Spirit: The Abbe de Condillac and the French Enlightenment.* New Haven: Yale University Press.

- R. L. E. Schwarzenberger (2000), *The Language of Geometry*, published in *A Mathematical Spectrum Miscellany*, Applied Probability Trust.

- Alan Ford & F. David Peat (1988), *The Role of Language in Science*, Foundations of Physics Vol 18.

- Kay O'Halloran, *Mathematical Discourse: Language, Symbolism and Visual Images*, Continuum, 2004. ISBN 0826468578

3.9 External links

- What is Language
- *Mathematics and the Language of Nature* - essay by F. David Peat.
- Mathematical Words: Origins and Sources (John Aldrich, University of Southampton)
- Communicating in the Language of Mathematics by Dr. David Moursund
- Handbook of Mathematical Discourse by Charles Wells.

Chapter 4

Formal language

This article is about a technical term in mathematics and computer science. For related studies about natural languages, see Formal semantics (linguistics). For formal modes of speech in natural languages, see Register (sociolinguistics).

In mathematics, computer science, and linguistics, a **formal language** is a set of strings of symbols that may be con-

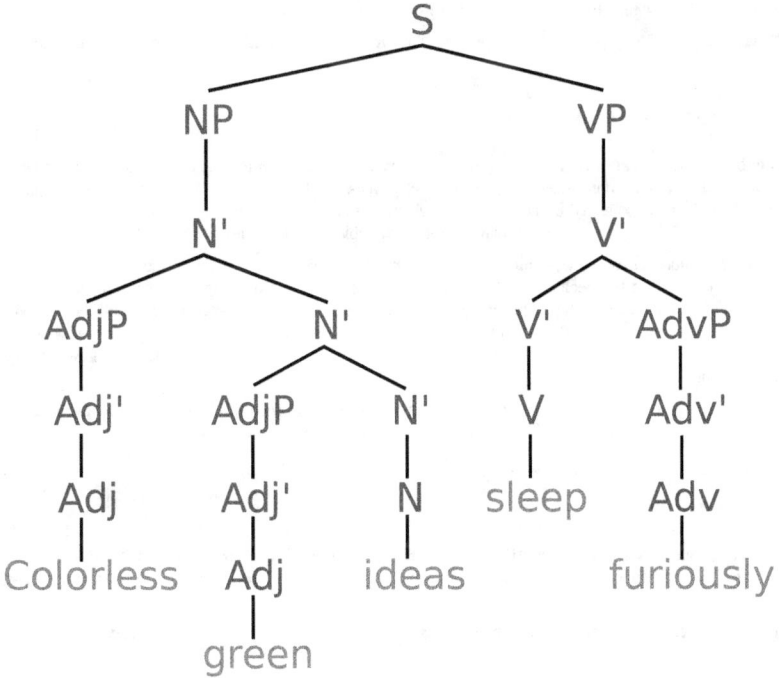

Structure of a syntactically well-formed, although nonsensical English sentence (historical example from Chomsky 1957).

strained by rules that are specific to it.

The alphabet of a formal language is the set of symbols, letters, or tokens from which the strings of the language may be formed; frequently it is required to be finite.[1] The strings formed from this alphabet are called words, and the words that belong to a particular formal language are sometimes called *well-formed words* or *well-formed formulas*. A formal language is often defined by means of a formal grammar such as a regular grammar or context-free grammar, also called its formation rule.

The field of **formal language theory** studies primarily the purely syntactical aspects of such languages—that is, their internal structural patterns. Formal language theory sprang out of linguistics, as a way of understanding the syntactic regularities of natural languages. In computer science, formal languages are used among others as the basis for defining the grammar of programming languages and formalized versions of subsets of natural languages in which the words of the language represent concepts that are associated with particular meanings or semantics. In computational complexity theory, decision problems are typically defined as formal languages, and complexity classes are defined as the sets of the formal languages that can be parsed by machines with limited computational power. In logic and the foundations of mathematics, formal languages are used to represent the syntax of axiomatic systems, and mathematical formalism is the philosophy that all of mathematics can be reduced to the syntactic manipulation of formal languages in this way.

4.1 History

The first formal language is thought be the one used by Gottlob Frege in his *Begriffsschrift* (1879), literally meaning "concept writing", and which Frege described as a "formal language of pure thought."[2]

Axel Thue's early Semi-Thue system which can be used for rewriting strings was influential on formal grammars.

4.2 Words over an alphabet

An **alphabet**, in the context of formal languages, can be any set, although it often makes sense to use an alphabet in the usual sense of the word, or more generally a character set such as ASCII or Unicode. Alphabets can also be infinite; e.g. first-order logic is often expressed using an alphabet which, besides symbols such as $\wedge$, $\neg$, $\vee$ and parentheses, contains infinitely many elements $x_0, x_1, x_2, \ldots$ that play the role of variables. The elements of an alphabet are called its **letters**.

A **word** over an alphabet can be any finite sequence, or string, of characters or letters, which sometimes may include spaces, and are separated by specified word separation characters. The set of all words over an alphabet Σ is usually denoted by Σ^* (using the Kleene star). The length of a word is the number of characters or letters it is composed of. For any alphabet there is only one word of length 0, the *empty word*, which is often denoted by e, ε or λ. By concatenation one can combine two words to form a new word, whose length is the sum of the lengths of the original words. The result of concatenating a word with the empty word is the original word.

In some applications, especially in logic, the alphabet is also known as the *vocabulary* and words are known as *formulas* or *sentences*; this breaks the letter/word metaphor and replaces it by a word/sentence metaphor.

4.3 Definition

A **formal language** L over an alphabet Σ is a subset of Σ^*, that is, a set of words over that alphabet. Sometimes the sets of words are grouped into expressions, whereas rules and constraints may be formulated for the creation of 'well-formed expressions'.

In computer science and mathematics, which do not usually deal with natural languages, the adjective "formal" is often omitted as redundant.

While formal language theory usually concerns itself with formal languages that are described by some syntactical rules, the actual definition of the concept "formal language" is only as above: a (possibly infinite) set of finite-length strings composed from a given alphabet, no more nor less. In practice, there are many languages that can be described by rules, such as regular languages or context-free languages. The notion of a formal grammar may be closer to the intuitive concept of a "language," one described by syntactic rules. By an abuse of the definition, a particular formal language is often thought of as being equipped with a formal grammar that describes it.

4.4   Examples

The following rules describe a formal language L over the alphabet $\Sigma = \{\ 0, 1, 2, 3, 4, 5, 6, 7, 8, 9, +, =\ \}$:

- Every nonempty string that does not contain "+" or "=" and does not start with "0" is in L.

- The string "0" is in L.

- A string containing "=" is in L if and only if there is exactly one "=", and it separates two valid strings of L.

- A string containing "+" but not "=" is in L if and only if every "+" in the string separates two valid strings of L.

- No string is in L other than those implied by the previous rules.

Under these rules, the string "23+4=555" is in L, but the string "=234=+" is not. This formal language expresses natural numbers, well-formed addition statements, and well-formed addition equalities, but it expresses only what they look like (their syntax), not what they mean (semantics). For instance, nowhere in these rules is there any indication that "0" means the number zero, or that "+" means addition.

4.4.1   Constructions

For finite languages one can explicitly enumerate all well-formed words. For example, we can describe a language L as just L = {"a", "b", "ab", "cba"}. The degenerate case of this construction is the **empty language**, which contains no words at all (L = ∅).

However, even over a finite (non-empty) alphabet such as $\Sigma = \{a, b\}$ there are infinitely many words: "a", "abb", "ababba", "aaababbbbaab", ... Therefore formal languages are typically infinite, and describing an infinite formal language is not as simple as writing $L = \{$"a", "b", "ab", "cba"$\}$. Here are some examples of formal languages:

- $L = \Sigma^*$, the set of *all* words over Σ;

- $L = \{$"a"$\}^* = \{$"a"$^n\}$, where n ranges over the natural numbers and "a"^{n} means "a" repeated n times (this is the set of words consisting only of the symbol "a");

- the set of syntactically correct programs in a given programming language (the syntax of which is usually defined by a context-free grammar);

- the set of inputs upon which a certain Turing machine halts; or

- the set of maximal strings of alphanumeric ASCII characters on this line, i.e., the set {"the", "set", "of", "maximal", "strings", "alphanumeric", "ASCII", "characters", "on", "this", "line", "i", "e"}.

4.5   Language-specification formalisms

Formal language theory rarely concerns itself with particular languages (except as examples), but is mainly concerned with the study of various types of formalisms to describe languages. For instance, a language can be given as

- those strings generated by some formal grammar;

- those strings described or matched by a particular regular expression;

- those strings accepted by some automaton, such as a Turing machine or finite state automaton;

- those strings for which some decision procedure (an algorithm that asks a sequence of related YES/NO questions) produces the answer YES.

Typical questions asked about such formalisms include:

- What is their expressive power? (Can formalism X describe every language that formalism Y can describe? Can it describe other languages?)

- What is their recognizability? (How difficult is it to decide whether a given word belongs to a language described by formalism X?)

- What is their comparability? (How difficult is it to decide whether two languages, one described in formalism X and one in formalism Y, or in X again, are actually the same language?).

Surprisingly often, the answer to these decision problems is "it cannot be done at all", or "it is extremely expensive" (with a characterization of how expensive). Therefore, formal language theory is a major application area of computability theory and complexity theory. Formal languages may be classified in the Chomsky hierarchy based on the expressive power of their generative grammar as well as the complexity of their recognizing automaton. Context-free grammars and regular grammars provide a good compromise between expressivity and ease of parsing, and are widely used in practical applications.

4.6 Operations on languages

Certain operations on languages are common. This includes the standard set operations, such as union, intersection, and complement. Another class of operation is the element-wise application of string operations.

Examples: suppose L_1 and L_2 are languages over some common alphabet.

- The *concatenation* $L_1 L_2$ consists of all strings of the form vw where v is a string from L_1 and w is a string from L_2.

- The *intersection* $L_1 \cap L_2$ of L_1 and L_2 consists of all strings which are contained in both languages

- The *complement* $\neg L$ of a language with respect to a given alphabet consists of all strings over the alphabet that are not in the language.

- The Kleene star: the language consisting of all words that are concatenations of 0 or more words in the original language;

- *Reversal*:

  - Let e be the empty word, then $e^R = e$, and
  - for each non-empty word $w = x_1 \ldots x_n$ over some alphabet, let $w^R = x_n \ldots x_1$,
  - then for a formal language L, $L^R = \{w^R \mid w \in L\}$.

- String homomorphism

Such string operations are used to investigate closure properties of classes of languages. A class of languages is closed under a particular operation when the operation, applied to languages in the class, always produces a language in the same class again. For instance, the context-free languages are known to be closed under union, concatenation, and intersection with regular languages, but not closed under intersection or complement. The theory of trios and abstract families of languages studies the most common closure properties of language families in their own right.[3]

4.7   Applications

4.7.1   Programming languages

Main articles: Syntax (programming languages) and Compiler compiler

A compiler usually has two distinct components. A lexical analyzer, generated by a tool like lex, identifies the tokens of the programming language grammar, e.g. identifiers or keywords, which are themselves expressed in a simpler formal language, usually by means of regular expressions. At the most basic conceptual level, a parser, usually generated by a parser generator like yacc, attempts to decide if the source program is valid, that is if it belongs to the programming language for which the compiler was built. Of course, compilers do more than just parse the source code—they usually translate it into some executable format. Because of this, a parser usually outputs more than a yes/no answer, typically an abstract syntax tree, which is used by subsequent stages of the compiler to eventually generate an executable containing machine code that runs directly on the hardware, or some intermediate code that requires a virtual machine to execute.

4.7.2   Formal theories, systems and proofs

Main articles: Theory (mathematical logic) and Formal system

In mathematical logic, a *formal theory* is a set of sentences expressed in a formal language.

A *formal system* (also called a *logical calculus*, or a *logical system*) consists of a formal language together with a deductive apparatus (also called a *deductive system*). The deductive apparatus may consist of a set of transformation rules which may be interpreted as valid rules of inference or a set of axioms, or have both. A formal system is used to derive one expression from one or more other expressions. Although a formal language can be identified with its formulas, a formal system cannot be likewise identified by its theorems. Two formal systems $\mathcal{FS}$ and $\mathcal{FS}'$ may have all the same theorems and yet differ in some significant proof-theoretic way (a formula A may be a syntactic consequence of a formula B in one but not another for instance).

A *formal proof* or *derivation* is a finite sequence of well-formed formulas (which may be interpreted as propositions) each of which is an axiom or follows from the preceding formulas in the sequence by a rule of inference. The last sentence in the sequence is a theorem of a formal system. Formal proofs are useful because their theorems can be interpreted as true propositions.

Interpretations and models

Main articles: Formal semantics (logic), Interpretation (logic) and Model theory

Formal languages are entirely syntactic in nature but may be given semantics that give meaning to the elements of the language. For instance, in mathematical logic, the set of possible formulas of a particular logic is a formal language, and an interpretation assigns a meaning to each of the formulas—usually, a truth value.

The study of interpretations of formal languages is called formal semantics. In mathematical logic, this is often done in terms of model theory. In model theory, the terms that occur in a formula are interpreted as mathematical structures, and fixed compositional interpretation rules determine how the truth value of the formula can be derived from the interpretation of its terms; a *model* for a formula is an interpretation of terms such that the formula becomes true.

4.8   See also

- Combinatorics on words
- Grammar framework

This diagram shows the syntactic divisions within a formal system. Strings of symbols may be broadly divided into nonsense and well-formed formulas. The set of well-formed formulas is divided into theorems and non-theorems.

- Formal method
- Mathematical notation
- Associative array
- String (computer science)

4.9 References

4.9.1 Citation footnotes

[1] See e.g. Reghizzi, Stefano Crespi (2009), *Formal Languages and Compilation*, Texts in Computer Science, Springer, p. 8, ISBN 9781848820500, An alphabet is a finite set.

[2] Martin Davis (1995). "Influences of Mathematical Logic on Computer Science". In Rolf Herken. *The universal Turing machine: a half-century survey*. Springer. p. 290. ISBN 978-3-211-82637-9.

[3] Hopcroft & Ullman (1979), Chapter 11: Closure properties of families of languages.

4.9.2 General references

- A. G. Hamilton, *Logic for Mathematicians*, Cambridge University Press, 1978, ISBN 0-521-21838-1.

- Seymour Ginsburg, *Algebraic and automata theoretic properties of formal languages*, North-Holland, 1975, ISBN 0-7204-2506-9.

- Michael A. Harrison, *Introduction to Formal Language Theory*, Addison-Wesley, 1978.

- John E. Hopcroft and Jeffrey D. Ullman, *Introduction to Automata Theory, Languages, and Computation*, Addison-Wesley Publishing, Reading Massachusetts, 1979. ISBN 81-7808-347-7.

- Rautenberg, Wolfgang (2010). *A Concise Introduction to Mathematical Logic* (3rd ed.). New York: Springer Science+Business Media. doi:10.1007/978-1-4419-1221-3. ISBN 978-1-4419-1220-6.

- Grzegorz Rozenberg, Arto Salomaa, *Handbook of Formal Languages: Volume I-III*, Springer, 1997, ISBN 3-540-61486-9.

- Patrick Suppes, *Introduction to Logic*, D. Van Nostrand, 1957, ISBN 0-442-08072-7.

4.10 External links

- Hazewinkel, Michiel, ed. (2001), "Formal language", *Encyclopedia of Mathematics*, Springer, ISBN 978-1-55608-010-4

- Alphabet at PlanetMath.org.

- Language at PlanetMath.org.

- University of Maryland, Formal Language Definitions

- James Power, "Notes on Formal Language Theory and Parsing", 29 November 2002.

- Drafts of some chapters in the "Handbook of Formal Language Theory", Vol. 1-3, G. Rozenberg and A. Salomaa (eds.), Springer Verlag, (1997):

  - Alexandru Mateescu and Arto Salomaa, "Preface" in Vol.1, pp. v-viii, and "Formal Languages: An Introduction and a Synopsis", Chapter 1 in Vol. 1, pp.1-39

  - Sheng Yu, "Regular Languages", Chapter 2 in Vol. 1

  - Jean-Michel Autebert, Jean Berstel, Luc Boasson, "Context-Free Languages and Push-Down Automata", Chapter 3 in Vol. 1

  - Christian Choffrut and Juhani Karhumäki, "Combinatorics of Words", Chapter 6 in Vol. 1

  - Tero Harju and Juhani Karhumäki, "Morphisms", Chapter 7 in Vol. 1, pp. 439 - 510

  - Jean-Eric Pin, "Syntactic semigroups", Chapter 10 in Vol. 1, pp. 679-746

  - M. Crochemore and C. Hancart, "Automata for matching patterns", Chapter 9 in Vol. 2

  - Dora Giammarresi, Antonio Restivo, "Two-dimensional Languages", Chapter 4 in Vol. 3, pp. 215 - 267

Chapter 5

Elementary proof

In mathematics, an **elementary proof** is a mathematical proof that only uses basic techniques. More specifically, the term is used in number theory to refer to proofs that make no use of complex analysis. For some time it was thought that certain theorems, like the prime number theorem, could only be proved using "higher" mathematics. However, over time, many of these results have been reproved using only elementary techniques.

While the meaning has not always been defined precisely, the term is commonly used in mathematical jargon. An elementary proof is not necessarily simple, in the sense of being easy to understand: some elementary proofs can be quite complicated.[1]

5.1 Prime number theorem

The distinction between elementary and non-elementary proofs has been considered especially important in regard to the prime number theorem. This theorem was first proved in 1896 by Jacques Hadamard and Charles Jean de la Vallée-Poussin using complex analysis. Many mathematicians then attempted to construct elementary proofs of the theorem, without success. G. H. Hardy expressed strong reservations; he considered that the essential "depth" of the result ruled out elementary proofs:

> No elementary proof of the prime number theorem is known, and one may ask whether it is reasonable to expect one. Now we know that the theorem is roughly equivalent to a theorem about an analytic function, the theorem that Riemann's zeta function has no roots on a certain line. A proof of such a theorem, not fundamentally dependent on the theory of functions, seems to me extraordinarily unlikely. It is rash to assert that a mathematical theorem *cannot* be proved in a particular way; but one thing seems quite clear. We have certain views about the logic of the theory; we think that some theorems, as we say "lie deep" and others nearer to the surface. If anyone produces an elementary proof of the prime number theorem, he will show that these views are wrong, that the subject does not hang together in the way we have supposed, and that it is time for the books to be cast aside and for the theory to be rewritten.
> —G. H. Hardy (1921). Lecture to Mathematical Society of Copenhagen. Quoted in Goldfeld (2003), p. 3

However, in 1948, Atle Selberg produced new methods which led him and Paul Erdős to find elementary proofs of the prime number theorem.[2]

A possible formalization of the notion of "elementary" in connection to a proof of a number-theoretical result is the restriction that the proof can be carried out in Peano arithmetic. Also in that sense, these proofs are elementary.

5.2 Friedman's conjecture

Main article: Grand conjecture

Harvey Friedman conjectured, "Every theorem published in the *Annals of Mathematics* whose statement involves only finitary mathematical objects (i.e., what logicians call an arithmetical statement) can be proved in elementary arithmetic."[3] The form of elementary arithmetic referred to in this conjecture can be formalized by a small set of axioms concerning integer arithmetic and mathematical induction. For instance, according to this conjecture, Fermat's Last Theorem should have an elementary proof; Wiles' proof of Fermat's Last Theorem is not elementary. However, there are other simple statements about arithmetic such as the existence of iterated exponential functions that cannot be proven in this theory.

5.3 References

[1] Diamond, Harold G. (1982), "Elementary methods in the study of the distribution of prime numbers", *Bulletin of the American Mathematical Society* **7** (3): 553–89, doi:10.1090/S0273-0979-1982-15057-1, MR 670132.

[2] Goldfeld, Dorian M. (2003), *The Elementary Proof of the Prime Number Theorem: An Historical Perspective* (PDF), p. 3, retrieved October 31, 2009

[3] Avigad, Jeremy (2003), "Number theory and elementary arithmetic" (PDF), *Philosophia Mathematica* **11** (3): 257, at 258, doi:10.1093/philmat/11.3.257.

Chapter 6

Formal proof

A **formal proof** or **derivation** is a finite sequence of sentences (called well-formed formulas in the case of a formal language) each of which is an axiom, an assumption, or follows from the preceding sentences in the sequence by a rule of inference. The last sentence in the sequence is a theorem of a formal system. The notion of theorem is not in general effective, therefore there may be no method by which we can always find a proof of a given sentence or determine that none exists. The concept of natural deduction is a generalization of the concept of proof.[1]

The theorem is a syntactic consequence of all the well-formed formulas preceding it in the proof. For a well-formed formula to qualify as part of a proof, it must be the result of applying a rule of the deductive apparatus of some formal system to the previous well-formed formulae in the proof sequence.

Formal proofs often are constructed with the help of computers in interactive theorem proving. Significantly, these proofs can be checked automatically, also by computer. Checking formal proofs is usually simple, while the problem of *finding* proofs (automated theorem proving) is usually computationally intractable and/or only semi-decidable, depending upon the formal system in use.

6.1 Background

6.1.1 Formal language

Main article: Formal language

A *formal language* is a set of finite sequences of symbols. Such a language can be defined without reference to any meanings of any of its expressions; it can exist before any interpretation is assigned to it – that is, before it has any meaning. Formal proofs are expressed in some formal language.

6.1.2 Formal grammar

Main articles: Formal grammar and Formation rule

A *formal grammar* (also called *formation rules*) is a precise description of the well-formed formulas of a formal language. It is synonymous with the set of strings over the alphabet of the formal language which constitute well formed formulas. However, it does not describe their semantics (i.e. what they mean).

6.1.3 Formal systems

Main article: Formal system

A *formal system* (also called a *logical calculus*, or a *logical system*) consists of a formal language together with a deductive apparatus (also called a *deductive system*). The deductive apparatus may consist of a set of transformation rules (also called *inference rules*) or a set of axioms, or have both. A formal system is used to derive one expression from one or more other expressions.

6.1.4 Interpretations

Main articles: Formal semantics (logic) and Interpretation (logic)

An *interpretation* of a formal system is the assignment of meanings to the symbols, and truth-values to the sentences of a formal system. The study of interpretations is called formal semantics. *Giving an interpretation* is synonymous with *constructing a model.*

6.2   See also

- Proof (truth)

- Mathematical proof

- Proof theory

- Axiomatic system

6.3   References

[1] The Cambridge Dictionary of Philosophy, *deduction*

6.4   External links

- "A Special Issue on Formal Proof". *Notices of the American Mathematical Society.* December 2008.

- 2πix.com: Logic Part of a series of articles covering mathematics and logic.

Chapter 7

Proof theory

Proof theory is a branch of mathematical logic that represents proofs as formal mathematical objects, facilitating their analysis by mathematical techniques. Proofs are typically presented as inductively-defined data structures such as plain lists, boxed lists, or trees, which are constructed according to the axioms and rules of inference of the logical system. As such, proof theory is syntactic in nature, in contrast to model theory, which is semantic in nature. Together with model theory, axiomatic set theory, and recursion theory, proof theory is one of the so-called *four pillars* of the foundations of mathematics.[1]

Proof theory is important in philosophical logic, where the primary interest is in the idea of a proof-theoretic semantics, an idea which depends upon technical ideas in structural proof theory to be feasible.

7.1 History

Although the formalisation of logic was much advanced by the work of such figures as Gottlob Frege, Giuseppe Peano, Bertrand Russell, and Richard Dedekind, the story of modern proof theory is often seen as being established by David Hilbert, who initiated what is called Hilbert's program in the foundations of mathematics. Kurt Gödel's seminal work on proof theory first advanced, then refuted this program: his completeness theorem initially seemed to bode well for Hilbert's aim of reducing all mathematics to a finitist formal system; then his incompleteness theorems showed that this is unattainable. All of this work was carried out with the proof calculi called the Hilbert systems.

In parallel, the foundations of structural proof theory were being founded. Jan Łukasiewicz suggested in 1926 that one could improve on Hilbert systems as a basis for the axiomatic presentation of logic if one allowed the drawing of conclusions from assumptions in the inference rules of the logic. In response to this Stanisław Jaśkowski (1929) and Gerhard Gentzen (1934) independently provided such systems, called calculi of natural deduction, with Gentzen's approach introducing the idea of symmetry between the grounds for asserting propositions, expressed in introduction rules, and the consequences of accepting propositions in the elimination rules, an idea that has proved very important in proof theory.[2] Gentzen (1934) further introduced the idea of the sequent calculus, a calculus advanced in a similar spirit that better expressed the duality of the logical connectives,[3] and went on to make fundamental advances in the formalisation of intuitionistic logic, and provide the first combinatorial proof of the consistency of Peano arithmetic. Together, the presentation of natural deduction and the sequent calculus introduced the fundamental idea of analytic proof to proof theory.

7.2 Formal and informal proof

Main article: Formal proof

The *informal* proofs of everyday mathematical practice are unlike the *formal* proofs of proof theory. They are rather like high-level sketches that would allow an expert to reconstruct a formal proof at least in principle, given enough time and

45

patience. For most mathematicians, writing a fully formal proof is too pedantic and long-winded to be in common use.

Formal proofs are constructed with the help of computers in interactive theorem proving. Significantly, these proofs can be checked automatically, also by computer. (Checking formal proofs is usually simple, whereas *finding* proofs (automated theorem proving) is generally hard.) An informal proof in the mathematics literature, by contrast, requires weeks of peer review to be checked, and may still contain errors.

7.3   Kinds of proof calculi

The three most well-known styles of proof calculi are:

- The Hilbert calculi

- The natural deduction calculi

- The sequent calculi

Each of these can give a complete and axiomatic formalization of propositional or predicate logic of either the classical or intuitionistic flavour, almost any modal logic, and many substructural logics, such as relevance logic or linear logic. Indeed it is unusual to find a logic that resists being represented in one of these calculi.

7.4   Consistency proofs

Main article: Consistency proof

As previously mentioned, the spur for the mathematical investigation of proofs in formal theories was Hilbert's program. The central idea of this program was that if we could give finitary proofs of consistency for all the sophisticated formal theories needed by mathematicians, then we could ground these theories by means of a metamathematical argument, which shows that all of their purely universal assertions (more technically their provable Π_1^0 sentences) are finitarily true; once so grounded we do not care about the non-finitary meaning of their existential theorems, regarding these as pseudo-meaningful stipulations of the existence of ideal entities.

The failure of the program was induced by Kurt Gödel's incompleteness theorems, which showed that any ω-consistent theory that is sufficiently strong to express certain simple arithmetic truths, cannot prove its own consistency, which on Gödel's formulation is a Π_1^0 sentence.

Much investigation has been carried out on this topic since, which has in particular led to:

- Refinement of Gödel's result, particularly J. Barkley Rosser's refinement, weakening the above requirement of ω-consistency to simple consistency;

- Axiomatisation of the core of Gödel's result in terms of a modal language, provability logic;

- Transfinite iteration of theories, due to Alan Turing and Solomon Feferman;

- The recent discovery of self-verifying theories, systems strong enough to talk about themselves, but too weak to carry out the diagonal argument that is the key to Gödel's unprovability argument.

7.5   Structural proof theory

Main article: Structural proof theory

Structural proof theory is the subdiscipline of proof theory that studies proof calculi that support a notion of analytic proof. The notion of analytic proof was introduced by Gentzen for the sequent calculus; there the analytic proofs are those that are cut-free. His natural deduction calculus also supports a notion of analytic proof, as shown by Dag Prawitz. The definition is slightly more complex: we say the analytic proofs are the normal forms, which are related to the notion of normal form in term rewriting. More exotic proof calculi such as Jean-Yves Girard's proof nets also support a notion of analytic proof.

Structural proof theory is connected to type theory by means of the Curry-Howard correspondence, which observes a structural analogy between the process of normalisation in the natural deduction calculus and beta reduction in the typed lambda calculus. This provides the foundation for the intuitionistic type theory developed by Per Martin-Löf, and is often extended to a three way correspondence, the third leg of which are the cartesian closed categories.

7.6 Proof-theoretic semantics

Main articles: proof-theoretic semantics and logical harmony

In linguistics, type-logical grammar, categorial grammar and Montague grammar apply formalisms based on structural proof theory to give a formal natural language semantics.

7.7 Tableau systems

Main article: Method of analytic tableaux

Analytic tableaux apply the central idea of analytic proof from structural proof theory to provide decision procedures and semi-decision procedures for a wide range of logics.

7.8 Ordinal analysis

Main article: Ordinal analysis

Ordinal analysis is a powerful technique for providing combinatorial consistency proofs for theories formalising arithmetic and analysis.

7.9 Logics from proof analysis

Main article: Substructural logic

Several important logics have come from insights into logical structure arising in structural proof theory.

7.10 See also

- Intermediate logic
- Model theory
- Proof (truth)

- Proof techniques

7.11   Notes

[1] E.g., Wang (1981), pp. 3–4, and Barwise (1978).

[2] Prawitz (2006, p. 98).

[3] Girard, Lafont, and Taylor (1988).

7.12   References

- J. Avigad, E.H. Reck (2001). "Clarifying the nature of the infinite": the development of metamathematics and proof theory. Carnegie-Mellon Technical Report CMU-PHIL-120.

- J. Barwise (ed., 1978). Handbook of Mathematical Logic. North-Holland.

- A. S. Troelstra, H. Schwichtenberg (1996). *Basic Proof Theory*. In series *Cambridge Tracts in Theoretical Computer Science*, Cambridge University Press, ISBN 0-521-77911-1.

- G. Gentzen (1935/1969). Investigations into logical deduction. In M. E. Szabo, editor, *Collected Papers of Gerhard Gentzen*. North-Holland. Translated by Szabo from "Untersuchungen über das logische Schliessen", Mathematisches Zeitschrift 39: 176-210, 405-431.

- Hazewinkel, Michiel, ed. (2001), "Proof theory", *Encyclopedia of Mathematics*, Springer, ISBN 978-1-55608-010-4

- Luis Moreno & Bharath Sriraman (2005).*Structural Stability and Dynamic Geometry: Some Ideas on Situated Proof. International Reviews on Mathematical Education. Vol. 37, no.3, pp. 130–139*

- Prawitz, Dag (2006) [1965]. *Natural deduction: A proof-theoretical study*. Mineola, New York: Dover Publications. ISBN 978-0-486-44655-4.

- J. von Plato (2008). The Development of Proof Theory. Stanford Encyclopedia of Philosophy.

- Wang, Hao (1981). *Popular Lectures on Mathematical Logic*. Van Nostrand Reinhold Company. ISBN 0-442-23109-1.

Chapter 8

Theorem

For the Italian film, see Teorema (film).

In mathematics, a **theorem** is a statement that has been proven on the basis of previously established statements, such as other theorems—and generally accepted statements, such as axioms. The proof of a mathematical theorem is a logical argument for the theorem statement given in accord with the rules of a deductive system. The proof of a theorem is often interpreted as justification of the truth of the theorem statement. In light of the requirement that theorems be proved, the concept of a theorem is fundamentally *deductive*, in contrast to the notion of a scientific theory, which is *empirical*.[2]

Many mathematical theorems are conditional statements. In this case, the proof deduces the conclusion from conditions called **hypotheses** or premises. In light of the interpretation of proof as justification of truth, the conclusion is often viewed as a necessary consequence of the hypotheses, namely, that the conclusion is true in case the hypotheses are true, without any further assumptions. However, the conditional could be interpreted differently in certain deductive systems, depending on the meanings assigned to the derivation rules and the conditional symbol.

Although they can be written in a completely symbolic form, for example, within the propositional calculus, theorems are often expressed in a natural language such as English. The same is true of proofs, which are often expressed as logically organized and clearly worded informal arguments, intended to convince readers of the truth of the statement of the theorem beyond any doubt, and from which a formal symbolic proof can in principle be constructed. Such arguments are typically easier to check than purely symbolic ones—indeed, many mathematicians would express a preference for a proof that not only demonstrates the validity of a theorem, but also explains in some way *why* it is obviously true. In some cases, a picture alone may be sufficient to prove a theorem. Because theorems lie at the core of mathematics, they are also central to its aesthetics. Theorems are often described as being "trivial", or "difficult", or "deep", or even "beautiful". These subjective judgments vary not only from person to person, but also with time: for example, as a proof is simplified or better understood, a theorem that was once difficult may become trivial. On the other hand, a deep theorem may be simply stated, but its proof may involve surprising and subtle connections between disparate areas of mathematics. Fermat's Last Theorem is a particularly well-known example of such a theorem.

8.1 Informal account of theorems

Logically, many theorems are of the form of an indicative conditional: *if A, then B*. Such a theorem does not assert *B*, only that *B* is a necessary consequence of *A*. In this case *A* is called the **hypothesis** of the theorem (note that "hypothesis" here is something very different from a conjecture) and *B* the **conclusion** (formally, *A* and *B* are termed the *antecedent* and *consequent*). The theorem "If *n* is an even natural number then *n*/2 is a natural number" is a typical example in which the hypothesis is "*n* is an even natural number" and the conclusion is "*n*/2 is also a natural number".

To be proven, a theorem must be expressible as a precise, formal statement. Nevertheless, theorems are usually expressed in natural language rather than in a completely symbolic form, with the intention that the reader can produce a formal statement from the informal one.

It is common in mathematics to choose a number of hypotheses within a given language and declare that the theory consists of all statements provable from these hypotheses. These hypothesis form the foundational basis of the theory and are called axioms or postulates. The field of mathematics known as proof theory studies formal languages, axioms and the structure of proofs.

Some theorems are "trivial", in the sense that they follow from definitions, axioms, and other theorems in obvious ways and do not contain any surprising insights. Some, on the other hand, may be called "deep", because their proofs may be long and difficult, involve areas of mathematics superficially distinct from the statement of the theorem itself, or show surprising connections between disparate areas of mathematics.[3] A theorem might be simple to state and yet be deep. An excellent example is Fermat's Last Theorem, and there are many other examples of simple yet deep theorems in number theory and combinatorics, among other areas.

Other theorems have a known proof that cannot easily be written down. The most prominent examples are the four color theorem and the Kepler conjecture. Both of these theorems are only known to be true by reducing them to a computational search that is then verified by a computer program. Initially, many mathematicians did not accept this form of proof, but it has become more widely accepted. The mathematician Doron Zeilberger has even gone so far as to claim that these are possibly the only nontrivial results that mathematicians have ever proved.[4] Many mathematical theorems can be reduced to more straightforward computation, including polynomial identities, trigonometric identities and hypergeometric identities.[5]

8.2   Provability and theoremhood

To establish a mathematical statement as a theorem, a proof is required, that is, a line of reasoning from axioms in the system (and other, already established theorems) to the given statement must be demonstrated. However, the proof is usually considered as separate from the theorem statement. Although more than one proof may be known for a single theorem, only one proof is required to establish the status of a statement as a theorem. The Pythagorean theorem and the law of quadratic reciprocity are contenders for the title of theorem with the greatest number of distinct proofs.

8.3   Relation with scientific theories

Theorems in mathematics and theories in science are fundamentally different in their epistemology. A scientific theory cannot be proven; its key attribute is that it is falsifiable, that is, it makes predictions about the natural world that are testable by experiments. Any disagreement between prediction and experiment demonstrates the incorrectness of the scientific theory, or at least limits its accuracy or domain of validity. Mathematical theorems, on the other hand, are purely abstract formal statements: the proof of a theorem cannot involve experiments or other empirical evidence in the same way such evidence is used to support scientific theories.

Nonetheless, there is some degree of empiricism and data collection involved in the discovery of mathematical theorems. By establishing a pattern, sometimes with the use of a powerful computer, mathematicians may have an idea of what to prove, and in some cases even a plan for how to set about doing the proof. For example, the Collatz conjecture has been verified for start values up to about 2.88×10^{18}. The Riemann hypothesis has been verified for the first 10 trillion zeroes of the zeta function. Neither of these statements is considered proven.

Such evidence does not constitute proof. For example, the Mertens conjecture is a statement about natural numbers that is now known to be false, but no explicit counterexample (i.e., a natural number n for which the Mertens function $M(n)$ equals or exceeds the square root of n) is known: all numbers less than 10^{14} have the Mertens property, and the smallest number that does not have this property is only known to be less than the exponential of 1.59×10^{40}, which is approximately 10 to the power 4.3×10^{39}. Since the number of particles in the universe is generally considered less than 10 to the power 100 (a googol), there is no hope to find an explicit counterexample by exhaustive search.

Note that the word "theory" also exists in mathematics, to denote a body of mathematical axioms, definitions and theorems, as in, for example, group theory. There are also "theorems" in science, particularly physics, and in engineering, but they often have statements and proofs in which physical assumptions and intuition play an important role; the physical axioms on which such "theorems" are based are themselves falsifiable.

8.4 Terminology

A number of different terms for mathematical statements exist, these terms indicate the role statements play in a particular subject. The distinction between different terms is sometimes rather arbitrary and the usage of some terms has evolved over time.

- An **axiom** or **postulate** is a statement that is accepted without proof and regarded as fundamental to a subject. Historically these have been regarded as "self-evident", but more recently they are considered assumptions that characterize the subject of study. In classical geometry, axioms are general statements, while postulates are statements about geometrical objects.[6] A definition is also accepted without proof since it simply gives the meaning of a word or phrase in terms of known concepts.

- An unproven statement that is believed true is called a **conjecture** (or sometimes a **hypothesis**, but with a different meaning from the one discussed above). To be considered a conjecture, a statement must usually be proposed publicly, at which point the name of the proponent may be attached to the conjecture, as with Goldbach's conjecture. Other famous conjectures include the Collatz conjecture and the Riemann hypothesis. On the other hand, Fermat's last theorem has always been known by that name, even before it was proven; it was never known as "Fermat's conjecture".

- A **proposition** is a theorem of no particular importance. This term sometimes connotes a statement with a simple proof, while the term **theorem** is usually reserved for the most important results or those with long or difficult proofs. In classical geometry, a proposition may be a construction that satisfies given requirements; for example, Proposition 1 in Book I of Euclid's elements is the construction of an equilateral triangle.[7]

- A **lemma** is a "helping theorem", a proposition with little applicability except that it forms part of the proof of a larger theorem. In some cases, as the relative importance of different theorems becomes more clear, what was once considered a lemma is now considered a theorem, though the word "lemma" remains in the name. Examples include Gauss's lemma, Zorn's lemma, and the Fundamental lemma.

- A **corollary** is a proposition that follows with little or no proof from another theorem or definition.[8]

- A **converse** of a theorem is a statement formed by interchanging what is given in a theorem and what is to be proved. For example, the isosceles triangle theorem states that if two sides of a triangle are equal then two angles are equal. In the converse, the given (that two sides are equal) and what is to be proved (that two angles are equal) are swapped, so the converse is the statement that if two angles of a triangle are equal then two sides are equal. In this example, the converse can be proven as another theorem, but this is often not the case. For example, the converse to the theorem that two right angles are equal angles is the statement that two equal angles must be right angles, and this is clearly not always the case.[9]

- A **generalization** is a theorem which includes a previously proven theorem as a special case and hence as a corollary.

There are other terms, less commonly used, that are conventionally attached to proven statements, so that certain theorems are referred to by historical or customary names. For example:

- An **identity** is an equality, contained in a theorem, between two mathematical expressions that holds regardless of what values are used for any variables or parameters appearing in the expressions. Examples include Euler's formula and Vandermonde's identity.

- A **rule** is a theorem, such as Bayes' rule and Cramer's rule, that establishes a useful formula.

- A **law** or a **principle** is a theorem that applies in a wide range of circumstances. Examples include the law of large numbers, the law of cosines, Kolmogorov's zero-one law, Harnack's principle, the least upper bound principle, and the pigeonhole principle.[10]

A few well-known theorems have even more idiosyncratic names. The **division algorithm** (see Euclidean division) is a theorem expressing the outcome of division in the natural numbers and more general rings. The **Bézout's identity** is a theorem asserting that the greatest common divisor of two numbers may be written as a linear combination of these numbers. The **Banach–Tarski paradox** is a theorem in measure theory that is paradoxical in the sense that it contradicts common intuitions about volume in three-dimensional space.

8.5   Layout

A theorem and its proof are typically laid out as follows:

>   **Theorem** (name of person who proved it and year of discovery, proof or publication).
>   *Statement of theorem (sometimes called the* proposition*).*
>   **Proof**.
>   *Description of proof.*
>   *End mark.*

The end of the proof may be signalled by the letters Q.E.D. (*quod erat demonstrandum*) or by one of the tombstone marks "□" or "∎" meaning "End of Proof", introduced by Paul Halmos following their usage in magazine articles.

The exact style depends on the author or publication. Many publications provide instructions or macros for typesetting in the house style.

It is common for a theorem to be preceded by definitions describing the exact meaning of the terms used in the theorem. It is also common for a theorem to be preceded by a number of propositions or lemmas which are then used in the proof. However, lemmas are sometimes embedded in the proof of a theorem, either with nested proofs, or with their proofs presented after the proof of the theorem.

Corollaries to a theorem are either presented between the theorem and the proof, or directly after the proof. Sometimes, corollaries have proofs of their own that explain why they follow from the theorem.

8.6   Lore

It has been estimated that over a quarter of a million theorems are proved every year.[11]

The well-known aphorism, "A mathematician is a device for turning coffee into theorems", is probably due to Alfréd Rényi, although it is often attributed to Rényi's colleague Paul Erdős (and Rényi may have been thinking of Erdős), who was famous for the many theorems he produced, the number of his collaborations, and his coffee drinking.[12]

The classification of finite simple groups is regarded by some to be the longest proof of a theorem. It comprises tens of thousands of pages in 500 journal articles by some 100 authors. These papers are together believed to give a complete proof, and several ongoing projects hope to shorten and simplify this proof.[13] Another theorem of this type is the Four color theorem whose computer generated proof is too long for a human to read. It is certainly the longest known proof of a theorem whose statement can be easily understood by a layman.

8.7   Theorems in logic

Logic, especially in the field of proof theory, considers theorems as statements (called **formulas** or **well formed formulas**) of a formal language. The statements of the language are strings of symbols and may be broadly divided into nonsense and well-formed formulas. A set of **deduction rules**, also called **transformation rules** or rules of inference, must be provided. These deduction rules tell exactly when a formula can be derived from a set of premises. The set of well-formed formulas may be broadly divided into theorems and non-theorems. However, according to Hofstadter, a formal system often simply defines all its well-formed formula as theorems.[14]

Different sets of derivation rules give rise to different interpretations of what it means for an expression to be a theorem. Some derivation rules and formal languages are intended to capture mathematical reasoning; the most common examples use first-order logic. Other deductive systems describe term rewriting, such as the reduction rules for λ calculus.

The definition of theorems as elements of a formal language allows for results in proof theory that study the structure of formal proofs and the structure of provable formulas. The most famous result is Gödel's incompleteness theorem; by representing theorems about basic number theory as expressions in a formal language, and then representing this language within number theory itself, Gödel constructed examples of statements that are neither provable nor disprovable from axiomatizations of number theory.

A theorem may be expressed in a formal language (or "formalized"). A formal theorem is the purely formal analogue of a theorem. In general, a formal theorem is a type of well-formed formula that satisfies certain logical and syntactic conditions. The notation S is often used to indicate that S is a theorem.

Formal theorems consist of formulas of a formal language and the transformation rules of a formal system. Specifically, a formal theorem is always the last formula of a derivation in some formal system each formula of which is a logical consequence of the formulas that came before it in the derivation. The initially accepted formulas in the derivation are called its **axioms**, and are the basis on which the theorem is derived. A set of theorems is called a **theory**.

What makes formal theorems useful and of interest is that they can be interpreted as true propositions and their derivations may be interpreted as a proof of the truth of the resulting expression. A set of formal theorems may be referred to as a **formal theory**. A theorem whose interpretation is a true statement about a formal system is called a **metatheorem**.

8.7.1 Syntax and semantics

Main articles: Syntax (logic) and Formal semantics (logic)

The concept of a formal theorem is fundamentally syntactic, in contrast to the notion of a *true proposition*, which introduces semantics. Different deductive systems can yield other interpretations, depending on the presumptions of the derivation rules (i.e. belief, justification or other modalities). The soundness of a formal system depends on whether or not all of its theorems are also validities. A validity is a formula that is true under any possible interpretation, e.g. in classical propositional logic validities are tautologies. A formal system is considered semantically complete when all of its tautologies are also theorems.

8.7.2 Derivation of a theorem

Main article: Formal proof

The notion of a theorem is very closely connected to its formal proof (also called a "derivation"). To illustrate how derivations are done, we will work in a very simplified formal system. Let us call ours $\mathcal{FS}$ Its alphabet consists only of two symbols { **A**, **B** } and its formation rule for formulas is:

$\mathcal{FS}$

The single axiom of $\mathcal{FS}$ is:

ABBA.

The only rule of inference (transformation rule) for $\mathcal{FS}$ is:

Any occurrence of "**A**" in a theorem may be replaced by an occurrence of the string "**AB**" and the result is a theorem.

Theorems in $\mathcal{FS}$ are defined as those formulae that have a derivation ending with that formula. For example

1. **ABBA** (Given as axiom)

2. **ABBBA** (by applying the transformation rule)

3. **ABBBAB** (by applying the transformation rule)

is a derivation. Therefore, "**ABBBAB**" is a theorem of $\mathcal{FS}$. The notion of truth (or falsity) cannot be applied to the formula "**ABBBAB**" until an interpretation is given to its symbols. Thus in this example, the formula does not yet represent a proposition, but is merely an empty abstraction.

Two metatheorems of $\mathcal{FS}$ are:

Every theorem begins with "**A**".

Every theorem has exactly two "**A**"s.

8.7.3   Interpretation of a formal theorem

Main article: Interpretation (logic)

8.7.4   Theorems and theories

Main articles: Theory and Theory (mathematical logic)

8.8   See also

- Inference

- List of theorems

- Toy theorem

- Metamath – a language for developing strictly formalized mathematical definitions and proofs accompanied by a proof checker for this language and a growing database of thousands of proved theorems

8.9   Notes

[1] For full text of 2nd edition of 1940, see Elisha Scott Loomis. "The Pythagorean proposition: its demonstrations analyzed and classified, and bibliography of sources for data of the four kinds of proofs" (PDF). *Education Resources Information Center*. Institute of Education Sciences (IES) of the U.S. Department of Education. Retrieved 2010-09-26. Originally published in 1940 and reprinted in 1968 by National Council of Teachers of Mathematics.

[2] However, both theorems and theories are investigations. See Heath 1897 Introduction, The terminology of Archimedes, p. clxxxii:"theorem (θεὼρνμα) from θεωρεῖν to investigate"

[3] Weisstein, Eric W., "Deep Theorem", *MathWorld*.

[4] Doron Zeilberger. "Opinion 51".

[5] Petkovsek et al. 1996.

[6] Wentworth, G.; Smith, D.E. (1913). "Art. 46, 47". *Plane Geometry*. Ginn & Co.

[7] Wentworth & Smith Art. 50

[8] Wentworth & Smith Art. 51

[9] Follows Wentworth & Smith Art. 79

[10] The word *law* can also refer to an axiom, a rule of inference, or, in probability theory, a probability distribution.

[11] Hoffman 1998, p. 204.

[12] Hoffman 1998, p. 7.

[13] An enormous theorem: the classification of finite simple groups, Richard Elwes, Plus Magazine, Issue 41 December 2006.

[14] Hofstadter 1980

8.10   References

- Heath, Sir Thomas Little (1897). *The works of Archimedes*. Dover. Retrieved 2009-11-15.

- Hoffman, P. (1998). *The Man Who Loved Only Numbers: The Story of Paul Erdős and the Search for Mathematical Truth*. Hyperion, New York. ISBN 1-85702-829-5.

- Hofstadter, Douglas (1979). *Gödel, Escher, Bach: An Eternal Golden Braid*. Basic Books.

- Hunter, Geofrfrey (1996) [1973]. *Metalogic: An Introduction to the Metatheory of Standard First Order Logic*. University of California Press. ISBN 0-520-02356-0.

- Mates, Benson (1972). *Elementary Logic*. Oxford University Press. ISBN 0-19-501491-X.

- Petkovsek, Marko; Wilf, Herbert; Zeilberger, Doron (1996). *A = B*. A.K. Peters, Wellesley, Massachusetts. ISBN 1-56881-063-6.

8.11   External links

- Weisstein, Eric W., "Theorem", *MathWorld*.

- Theorem of the Day

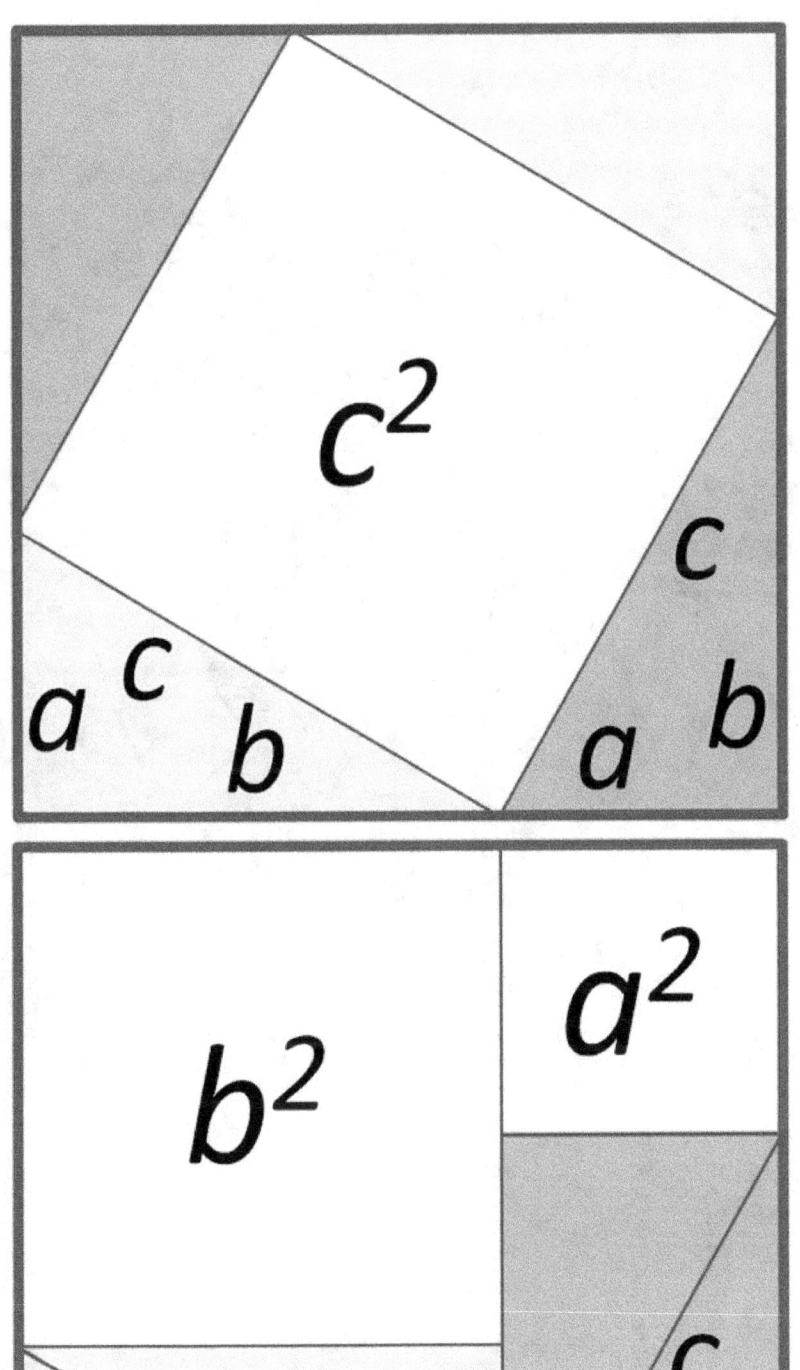

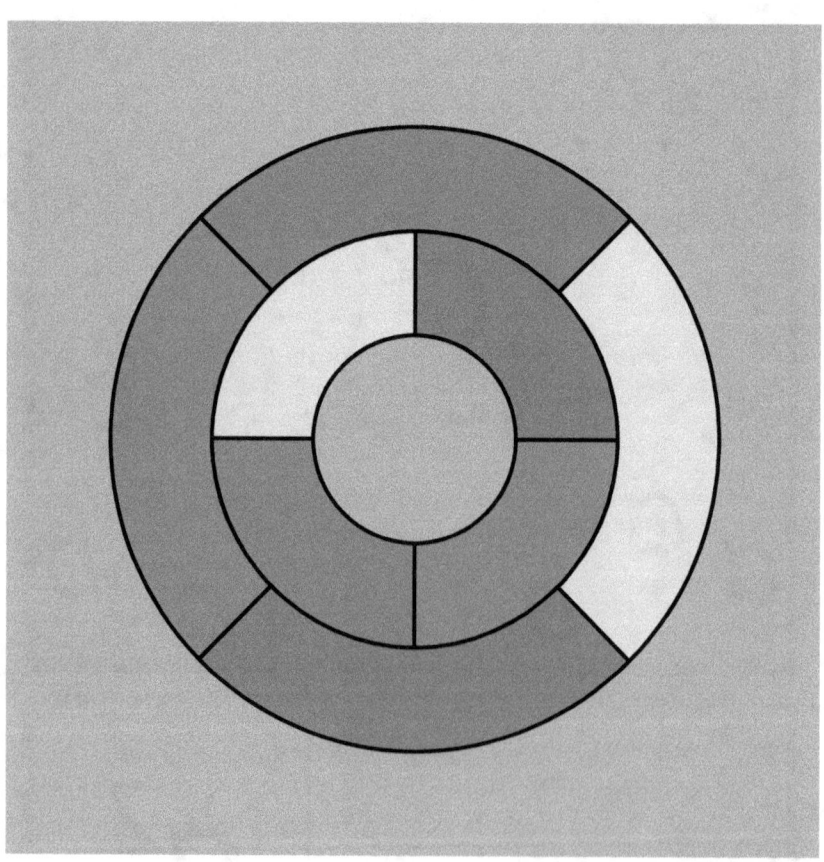

A planar map with five colors such that no two regions with the same color meet. It can actually be colored in this way with only four colors. The four color theorem states that such colorings are possible for any planar map, but every known proof involves a computational search that is too long to check by hand.

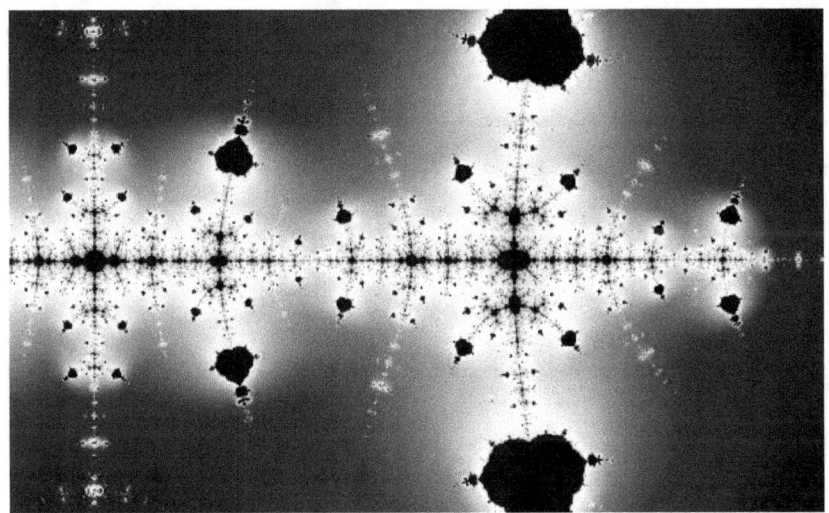

The Collatz conjecture: one way to illustrate its complexity is to extend the iteration from the natural numbers to the complex numbers. The result is a fractal, which (in accordance with universality) resembles the Mandelbrot set.

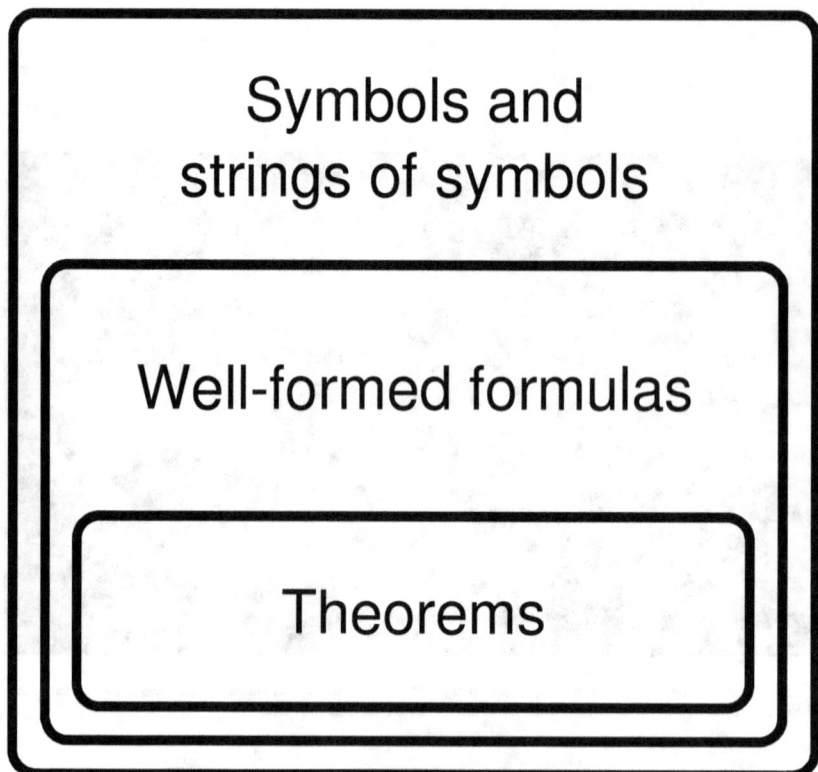

This diagram shows the syntactic entities that can be constructed from formal languages. The symbols and strings of symbols may be broadly divided into nonsense and well-formed formulas. A formal language can be thought of as identical to the set of its well-formed formulas. The set of well-formed formulas may be broadly divided into theorems and non-theorems.

Chapter 9

Axiom

This article is about logical propositions. For other uses, see Axiom (disambiguation).
"Axiomatic" redirects here. For other uses, see Axiomatic (disambiguation).
"Postulation" redirects here. For the term in algebraic geometry, see Postulation (algebraic geometry).

An **axiom** or **postulate** is a premise or starting point of reasoning. As classically conceived, an axiom is a premise so evident as to be accepted as true without controversy.[1] The word comes from the Greek *axíōma* (ἀξίωμα) 'that which is thought worthy or fit' or 'that which commends itself as evident.'[2][3] As used in modern logic, an axiom is simply a premise or starting point for reasoning.[4] What it means for an axiom, or any mathematical statement, to be "true" is a central question in the philosophy of mathematics, with modern mathematicians holding a multitude of different opinions.

In mathematics, the term *axiom* is used in two related but distinguishable senses: "logical axioms" and "non-logical axioms". Logical axioms are usually statements that are taken to be true within the system of logic they define (e.g., (*A* and *B*) implies *A*), while non-logical axioms (e.g., $a + b = b + a$) are actually substantive assertions about the elements of the domain of a specific mathematical theory (such as arithmetic). When used in the latter sense, "axiom," "postulate", and "assumption" may be used interchangeably. In general, a non-logical axiom is not a self-evident truth, but rather a formal logical expression used in deduction to build a mathematical theory. As modern mathematics admits multiple, equally "true" systems of logic, precisely the same thing must be said for logical axioms - they both define and are specific to the particular system of logic that is being invoked. To axiomatize a system of knowledge is to show that its claims can be derived from a small, well-understood set of sentences (the axioms). There are typically multiple ways to axiomatize a given mathematical domain.

In both senses, an axiom is any mathematical statement that serves as a starting point from which other statements are logically derived. Within the system they define, axioms (unless redundant) cannot be derived by principles of deduction, nor are they demonstrable by mathematical proofs, simply because they are starting points; there is nothing else from which they logically follow otherwise they would be classified as theorems. However, an axiom in one system may be a theorem in another, and vice versa.

9.1 Etymology

The word "axiom" comes from the Greek word ἀξίωμα (*axioma*), a verbal noun from the verb ἀξιόειν (*axioein*), meaning "to deem worthy", but also "to require", which in turn comes from ἄξιος (*axios*), meaning "being in balance", and hence "having (the same) value (as)", "worthy", "proper". Among the ancient Greek philosophers an axiom was a claim which could be seen to be true without any need for proof.

The root meaning of the word 'postulate' is to 'demand'; for instance, Euclid demands of us that we agree that some things can be done, e.g. any two points can be joined by a straight line, etc.[5]

Ancient geometers maintained some distinction between axioms and postulates. While commenting Euclid's books

Proclus remarks that "Geminus held that this [4th] Postulate should not be classed as a postulate but as an axiom, since it does not, like the first three Postulates, assert the possibility of some construction but expresses an essential property".[6] Boethius translated 'postulate' as *petitio* and called the axioms *notiones communes* but in later manuscripts this usage was not always strictly kept.

9.2 Historical development

9.2.1 Early Greeks

The logico-deductive method whereby conclusions (new knowledge) follow from premises (old knowledge) through the application of sound arguments (syllogisms, rules of inference), was developed by the ancient Greeks, and has become the core principle of modern mathematics. Tautologies excluded, nothing can be deduced if nothing is assumed. Axioms and postulates are the basic assumptions underlying a given body of deductive knowledge. They are accepted without demonstration. All other assertions (theorems, if we are talking about mathematics) must be proven with the aid of these basic assumptions. However, the interpretation of mathematical knowledge has changed from ancient times to the modern, and consequently the terms *axiom* and *postulate* hold a slightly different meaning for the present day mathematician, than they did for Aristotle and Euclid.

The ancient Greeks considered geometry as just one of several sciences, and held the theorems of geometry on par with scientific facts. As such, they developed and used the logico-deductive method as a means of avoiding error, and for structuring and communicating knowledge. Aristotle's posterior analytics is a definitive exposition of the classical view.

An "axiom", in classical terminology, referred to a self-evident assumption common to many branches of science. A good example would be the assertion that

> *When an equal amount is taken from equals, an equal amount results.*

At the foundation of the various sciences lay certain additional hypotheses which were accepted without proof. Such a hypothesis was termed a *postulate*. While the axioms were common to many sciences, the postulates of each particular science were different. Their validity had to be established by means of real-world experience. Indeed, Aristotle warns that the content of a science cannot be successfully communicated, if the learner is in doubt about the truth of the postulates.[7]

The classical approach is well-illustrated by Euclid's Elements, where a list of postulates is given (common-sensical geometric facts drawn from our experience), followed by a list of "common notions" (very basic, self-evident assertions).

Postulates

1. It is possible to draw a straight line from any point to any other point.
2. It is possible to extend a line segment continuously in both directions.
3. It is possible to describe a circle with any center and any radius.
4. It is true that all right angles are equal to one another.
5. ("Parallel postulate") It is true that, if a straight line falling on two straight lines make the interior angles on the same side less than two right angles, the two straight lines, if produced indefinitely, intersect on that side on which are the angles less than the two right angles.

Common notions

1. Things which are equal to the same thing are also equal to one another.
2. If equals are added to equals, the wholes are equal.
3. If equals are subtracted from equals, the remainders are equal.
4. Things which coincide with one another are equal to one another.
5. The whole is greater than the part.

9.2.2   Modern development

A lesson learned by mathematics in the last 150 years is that it is useful to strip the meaning away from the mathematical assertions (axioms, postulates, propositions, theorems) and definitions. One must concede the need for primitive notions, or undefined terms or concepts, in any study. Such abstraction or formalization makes mathematical knowledge more general, capable of multiple different meanings, and therefore useful in multiple contexts. Alessandro Padoa, Mario Pieri, and Giuseppe Peano were pioneers in this movement.

Structuralist mathematics goes further, and develops theories and axioms (e.g. field theory, group theory, topology, vector spaces) without *any* particular application in mind. The distinction between an "axiom" and a "postulate" disappears. The postulates of Euclid are profitably motivated by saying that they lead to a great wealth of geometric facts. The truth of these complicated facts rests on the acceptance of the basic hypotheses. However, by throwing out Euclid's fifth postulate we get theories that have meaning in wider contexts, hyperbolic geometry for example. We must simply be prepared to use labels like "line" and "parallel" with greater flexibility. The development of hyperbolic geometry taught mathematicians that postulates should be regarded as purely formal statements, and not as facts based on experience.

When mathematicians employ the field axioms, the intentions are even more abstract. The propositions of field theory do not concern any one particular application; the mathematician now works in complete abstraction. There are many examples of fields; field theory gives correct knowledge about them all.

It is not correct to say that the axioms of field theory are "propositions that are regarded as true without proof." Rather, the field axioms are a set of constraints. If any given system of addition and multiplication satisfies these constraints, then one is in a position to instantly know a great deal of extra information about this system.

Modern mathematics formalizes its foundations to such an extent that mathematical theories can be regarded as mathematical objects, and mathematics itself can be regarded as a branch of logic. Frege, Russell, Poincaré, Hilbert, and Gödel are some of the key figures in this development.

In the modern understanding, a set of axioms is any collection of formally stated assertions from which other formally stated assertions follow by the application of certain well-defined rules. In this view, logic becomes just another formal system. A set of axioms should be consistent; it should be impossible to derive a contradiction from the axiom. A set of axioms should also be non-redundant; an assertion that can be deduced from other axioms need not be regarded as an axiom.

It was the early hope of modern logicians that various branches of mathematics, perhaps all of mathematics, could be derived from a consistent collection of basic axioms. An early success of the formalist program was Hilbert's formalization of Euclidean geometry, and the related demonstration of the consistency of those axioms.

In a wider context, there was an attempt to base all of mathematics on Cantor's set theory. Here the emergence of Russell's paradox, and similar antinomies of naïve set theory raised the possibility that any such system could turn out to be inconsistent.

The formalist project suffered a decisive setback, when in 1931 Gödel showed that it is possible, for any sufficiently large set of axioms (Peano's axioms, for example) to construct a statement whose truth is independent of that set of axioms. As a corollary, Gödel proved that the consistency of a theory like Peano arithmetic is an unprovable assertion within the scope of that theory.

It is reasonable to believe in the consistency of Peano arithmetic because it is satisfied by the system of natural numbers, an infinite but intuitively accessible formal system. However, at present, there is no known way of demonstrating the consistency of the modern Zermelo–Fraenkel axioms for set theory. Furthermore, using techniques of forcing (Cohen) one can show that the continuum hypothesis (Cantor) is independent of the Zermelo–Fraenkel axioms. Thus, even this very general set of axioms cannot be regarded as the definitive foundation for mathematics.

9.2.3   Other sciences

Axioms play a key role not only in mathematics, but also in other sciences, notably in theoretical physics. In particular, the monumental work of Isaac Newton is essentially based on Euclid's axioms, augmented by a postulate on the non-relation of spacetime and the physics taking place in it at any moment.

In 1905, Newton's axioms were replaced by those of Albert Einstein's special relativity, and later on by those of general relativity.

Another paper of Albert Einstein and coworkers (see EPR paradox), almost immediately contradicted by Niels Bohr, concerned the interpretation of quantum mechanics. This was in 1935. According to Bohr, this new theory should be probabilistic, whereas according to Einstein it should be deterministic. Notably, the underlying quantum mechanical theory, i.e. the set of "theorems" derived by it, seemed to be identical. Einstein even assumed that it would be sufficient to add to quantum mechanics "hidden variables" to enforce determinism. However, thirty years later, in 1964, John Bell found a theorem, involving complicated optical correlations (see Bell inequalities), which yielded measurably different results using Einstein's axioms compared to using Bohr's axioms. And it took roughly another twenty years until an experiment of Alain Aspect got results in favour of Bohr's axioms, not Einstein's. (Bohr's axioms are simply: The theory should be probabilistic in the sense of the Copenhagen interpretation.)

As a consequence, it is not necessary to explicitly cite Einstein's axioms, the more so since they concern subtle points on the "reality" and "locality" of experiments.

Regardless, the role of axioms in mathematics and in the above-mentioned sciences is different. In mathematics one neither "proves" nor "disproves" an axiom for a set of theorems; the point is simply that in the conceptual realm identified by the axioms, the theorems logically follow. In contrast, in physics a comparison with experiments always makes sense, since a falsified physical theory needs modification.

9.3 Mathematical logic

In the field of mathematical logic, a clear distinction is made between two notions of axioms: *logical* and *non-logical* (somewhat similar to the ancient distinction between "axioms" and "postulates" respectively).

9.3.1 Logical axioms

These are certain formulas in a formal language that are universally valid, that is, formulas that are satisfied by every assignment of values. Usually one takes as logical axioms *at least* some minimal set of tautologies that is sufficient for proving all tautologies in the language; in the case of predicate logic more logical axioms than that are required, in order to prove logical truths that are not tautologies in the strict sense.

Examples

Propositional logic   In propositional logic it is common to take as logical axioms all formulae of the following forms, where ϕ, χ, and ψ can be any formulae of the language and where the included primitive connectives are only " $\neg$ " for negation of the immediately following proposition and " $\rightarrow$ " for implication from antecedent to consequent propositions:

1. $\phi \rightarrow (\psi \rightarrow \phi)$

2. $(\phi \rightarrow (\psi \rightarrow \chi)) \rightarrow ((\phi \rightarrow \psi) \rightarrow (\phi \rightarrow \chi))$

3. $(\neg\phi \rightarrow \neg\psi) \rightarrow (\psi \rightarrow \phi)$.

Each of these patterns is an *axiom schema*, a rule for generating an infinite number of axioms. For example, if A, B, and C are propositional variables, then $A \rightarrow (B \rightarrow A)$ and $(A \rightarrow \neg B) \rightarrow (C \rightarrow (A \rightarrow \neg B))$ are both instances of axiom schema 1, and hence are axioms. It can be shown that with only these three axiom schemata and *modus ponens*, one can prove all tautologies of the propositional calculus. It can also be shown that no pair of these schemata is sufficient for proving all tautologies with *modus ponens*.

Other axiom schemas involving the same or different sets of primitive connectives can be alternatively constructed.[8]

These axiom schemata are also used in the predicate calculus, but additional logical axioms are needed to include a quantifier in the calculus.[9]

First-order logic   Axiom of Equality. Let $\mathfrak{L}$ be a first-order language. For each variable x , the formula

$$x = x$$

is universally valid.

This means that, for any variable symbol x , the formula $x = x$ can be regarded as an axiom. Also, in this example, for this not to fall into vagueness and a never-ending series of "primitive notions", either a precise notion of what we mean by $x = x$ (or, for that matter, "to be equal") has to be well established first, or a purely formal and syntactical usage of the symbol $=$ has to be enforced, only regarding it as a string and only a string of symbols, and mathematical logic does indeed do that.

Another, more interesting example axiom scheme, is that which provides us with what is known as **Universal Instanti-ation**:

Axiom scheme for Universal Instantiation. Given a formula ϕ in a first-order language $\mathfrak{L}$, a variable x and a term t that is substitutable for x in ϕ , the formula

$$\forall x\, \phi \to \phi_t^x$$

is universally valid.

Where the symbol ϕ_t^x stands for the formula ϕ with the term t substituted for x . (See Substitution of variables.) In informal terms, this example allows us to state that, if we know that a certain property P holds for every x and that t stands for a particular object in our structure, then we should be able to claim $P(t)$. Again, *we are claiming that the formula $\forall x \phi \to \phi_t^x$ is valid*, that is, we must be able to give a "proof" of this fact, or more properly speaking, a *metaproof*. Actually, these examples are *metatheorems* of our theory of mathematical logic since we are dealing with the very concept of *proof* itself. Aside from this, we can also have **Existential Generalization**:

Axiom scheme for Existential Generalization. Given a formula ϕ in a first-order language $\mathfrak{L}$, a variable x and a term t that is substitutable for x in ϕ , the formula

$$\phi_t^x \to \exists x\, \phi$$

is universally valid.

9.3.2   Non-logical axioms

Non-logical axioms are formulas that play the role of theory-specific assumptions. Reasoning about two different struc-tures, for example the natural numbers and the integers, may involve the same logical axioms; the non-logical axioms aim to capture what is special about a particular structure (or set of structures, such as groups). Thus non-logical axioms, unlike logical axioms, are not *tautologies*. Another name for a non-logical axiom is *postulate*.[10]

Almost every modern mathematical theory starts from a given set of non-logical axioms, and it was thought that in principle every theory could be axiomatized in this way and formalized down to the bare language of logical formulas.

Non-logical axioms are often simply referred to as *axioms* in mathematical discourse. This does not mean that it is claimed that they are true in some absolute sense. For example, in some groups, the group operation is commutative, and this can be asserted with the introduction of an additional axiom, but without this axiom we can do quite well developing (the more general) group theory, and we can even take its negation as an axiom for the study of non-commutative groups.

Thus, an *axiom* is an elementary basis for a formal logic system that together with the rules of inference define a **deductive system**.

Examples

This section gives examples of mathematical theories that are developed entirely from a set of non-logical axioms (axioms, henceforth). A rigorous treatment of any of these topics begins with a specification of these axioms.

Basic theories, such as arithmetic, real analysis and complex analysis are often introduced non-axiomatically, but implicitly or explicitly there is generally an assumption that the axioms being used are the axioms of Zermelo–Fraenkel set theory with choice, abbreviated ZFC, or some very similar system of axiomatic set theory like Von Neumann–Bernays–Gödel set theory, a conservative extension of ZFC. Sometimes slightly stronger theories such as Morse-Kelley set theory or set theory with a strongly inaccessible cardinal allowing the use of a Grothendieck universe are used, but in fact most mathematicians can actually prove all they need in systems weaker than ZFC, such as second-order arithmetic.

The study of topology in mathematics extends all over through point set topology, algebraic topology, differential topology, and all the related paraphernalia, such as homology theory, homotopy theory. The development of *abstract algebra* brought with itself group theory, rings, fields, and Galois theory.

This list could be expanded to include most fields of mathematics, including measure theory, ergodic theory, probability, representation theory, and differential geometry.

Combinatorics is an example of a field of mathematics which does not, in general, follow the axiomatic method.

Arithmetic The Peano axioms are the most widely used *axiomatization* of first-order arithmetic. They are a set of axioms strong enough to prove many important facts about number theory and they allowed Gödel to establish his famous second incompleteness theorem.[11]

We have a language $\mathcal{L}_{NT} = \{0, S\}$ where 0 is a constant symbol and S is a unary function and the following axioms:

1. $\forall x. \neg (Sx = 0)$

2. $\forall x. \forall y. (Sx = Sy \rightarrow x = y)$

3. $((\phi(0) \wedge \forall x. (\phi(x) \rightarrow \phi(Sx))) \rightarrow \forall x. \phi(x)$ for any $\mathcal{L}_{NT}$ formula ϕ with one free variable.

The standard structure is $\mathfrak{N} = \langle \mathbb{N}, 0, S \rangle$ where $\mathbb{N}$ is the set of natural numbers, S is the successor function and 0 is naturally interpreted as the number 0.

Euclidean geometry Probably the oldest, and most famous, list of axioms are the 4 + 1 Euclid's postulates of plane geometry. The axioms are referred to as "4 + 1" because for nearly two millennia the fifth (parallel) postulate ("through a point outside a line there is exactly one parallel") was suspected of being derivable from the first four. Ultimately, the fifth postulate was found to be independent of the first four. Indeed, one can assume that exactly one parallel through a point outside a line exists, or that infinitely many exist. This choice gives us two alternative forms of geometry in which the interior angles of a triangle add up to exactly 180 degrees or less, respectively, and are known as Euclidean and hyperbolic geometries. If one also removes the second postulate ("a line can be extended indefinitely") then elliptic geometry arises, where there is no parallel through a point outside a line, and in which the interior angles of a triangle add up to more than 180 degrees.

Real analysis The object of study is the real numbers. The real numbers are uniquely picked out (up to isomorphism) by the properties of a *Dedekind complete ordered field*, meaning that any nonempty set of real numbers with an upper bound has a least upper bound. However, expressing these properties as axioms requires use of second-order logic. The Löwenheim-Skolem theorems tell us that if we restrict ourselves to first-order logic, any axiom system for the reals admits other models, including both models that are smaller than the reals and models that are larger. Some of the latter are studied in non-standard analysis.

9.3.3 Role in mathematical logic

Deductive systems and completeness

A **deductive system** consists of a set Λ of logical axioms, a set Σ of non-logical axioms, and a set $\{(\Gamma, \phi)\}$ of *rules of inference*. A desirable property of a deductive system is that it be **complete**. A system is said to be complete if, for all formulas ϕ,

$$\text{if } \Sigma \models \phi \text{ then } \Sigma \vdash \phi$$

that is, for any statement that is a *logical consequence* of Σ there actually exists a *deduction* of the statement from Σ . This is sometimes expressed as "everything that is true is provable", but it must be understood that "true" here means "made true by the set of axioms", and not, for example, "true in the intended interpretation". Gödel's completeness theorem establishes the completeness of a certain commonly used type of deductive system.

Note that "completeness" has a different meaning here than it does in the context of Gödel's first incompleteness theorem, which states that no *recursive, consistent* set of non-logical axioms Σ of the Theory of Arithmetic is *complete*, in the sense that there will always exist an arithmetic statement ϕ such that neither ϕ nor $\neg \phi$ can be proved from the given set of axioms.

There is thus, on the one hand, the notion of *completeness of a deductive system* and on the other hand that of *completeness of a set of non-logical axioms*. The completeness theorem and the incompleteness theorem, despite their names, do not contradict one another.

9.3.4 Further discussion

Early mathematicians regarded axiomatic geometry as a model of physical space, and obviously there could only be one such model. The idea that alternative mathematical systems might exist was very troubling to mathematicians of the 19th century and the developers of systems such as Boolean algebra made elaborate efforts to derive them from traditional arithmetic. Galois showed just before his untimely death that these efforts were largely wasted. Ultimately, the abstract parallels between algebraic systems were seen to be more important than the details and modern algebra was born. In the modern view axioms may be any set of formulas, as long as they are not known to be inconsistent.

9.4   See also

- Axiomatic system

- Dogma

- List of axioms

- Model theory

- Regulæ Juris

- Theorem

9.5   References

[1] "A proposition that commends itself to general acceptance; a well-established or universally conceded principle; a maxim, rule, law" axiom, n., definition 1a. *Oxford English Dictionary* Online, accessed 2012-04-28. Cf. Aristotle, *Posterior Analytics* I.2.72a18-b4.

[2] Cf. axiom, n., etymology. *Oxford English Dictionary*, accessed 2012-04-28.

[3] Oxford American College Dictionary: "n. a statement or proposition that is regarded as being established, accepted, or self-evidently true. ORIGIN: late 15th cent.: ultimately from Greek axiôma 'what is thought fitting,' from axios 'worthy.' http://www.highbeam.com/doc/1O997-axiom.html (subscription required)

[4] "A proposition (whether true or false)" axiom, n., definition 2. *Oxford English Dictionary* Online, accessed 2012-04-28.

[5] Wolff, P. Breakthroughs in Mathematics, 1963, New York: New American Library, pp 47–8

[6] Heath, T. 1956. The Thirteen Books of Euclid's Elements. New York: Dover. *p200*

[7] Aristotle, Metaphysics Bk IV, Chapter 3, 1005b "Physics also is a kind of Wisdom, but it is not the first kind. – And the attempts of some of those who discuss the terms on which truth should be accepted, are due to want of training in logic; for they should know these things already when they come to a special study, and not be inquiring into them while they are listening to lectures on it." W.D. Ross translation, in The Basic Works of Aristotle, ed. Richard McKeon, (Random House, New York, 1941)|date=June 2011

[8] Mendelson, "6. Other Axiomatizations" of Ch. 1

[9] Mendelson, "3. First-Order Theories" of Ch. 2

[10] Mendelson, "3. First-Order Theories: Proper Axioms" of Ch. 2

[11] Mendelson, "5. The Fixed Point Theorem. Gödel's Incompleteness Theorem" of Ch. 2

9.6 Further reading

- Mendelson, Elliot (1987). *Introduction to mathematical logic.* Belmont, California: Wadsworth & Brooks. ISBN 0-534-06624-0

9.7 External links

- Axiom at PhilPapers

- Axiom at PlanetMath.org.

- *Metamath* axioms page

Chapter 10

Conjecture

For text reconstruction, see Conjecture (textual criticism). For the annual science fiction convention held in San Diego, see Conjecture (convention).

In mathematics, a **conjecture** is a conclusion or proposition which appears to be correct based on incomplete information,

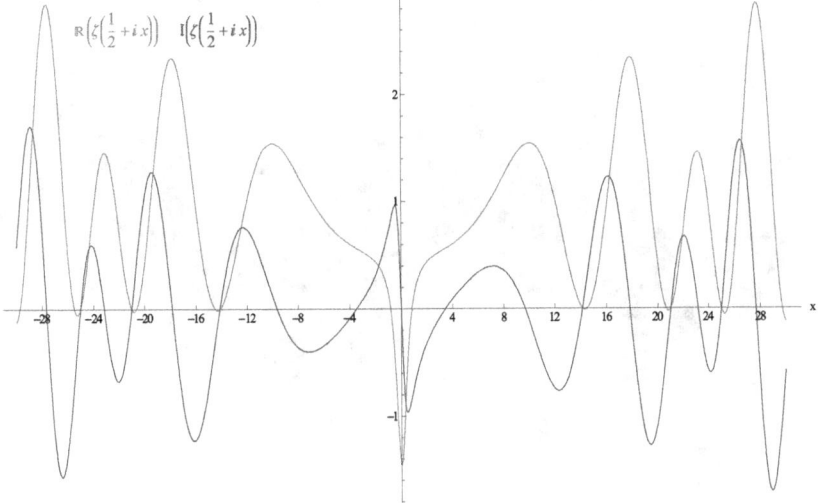

The real part (red) and imaginary part (blue) of the Riemann zeta function along the critical line Re(s) = 1/2. The first non-trivial zeros can be seen at Im(s) = ±14.135, ±21.022 and ±25.011. The Riemann hypothesis, a famous conjecture, says that all non-trivial zeros of the zeta function lie along the critical line.

but for which no proof has been found.[1][2] Conjectures such as the Riemann hypothesis or Fermat's Last Theorem have shaped much of mathematical history as new areas of mathematics are developed in order to solve them.

10.1 Important examples

10.1.1 Fermat's Last Theorem

Main article: Fermat's last theorem

In number theory, Fermat's Last Theorem (sometimes called **Fermat's conjecture**, especially in older texts) states that no three positive integers a, b, and c can satisfy the equation $a^n + b^n = c^n$ for any integer value of n greater than two.

This theorem was first conjectured by Pierre de Fermat in 1637 in the margin of a copy of *Arithmetica* where he claimed he had a proof that was too large to fit in the margin.[3] The first successful proof was released in 1994 by Andrew Wiles, and formally published in 1995, after 358 years of effort by mathematicians. The unsolved problem stimulated the development of algebraic number theory in the 19th century and the proof of the modularity theorem in the 20th century. It is among the most notable theorems in the history of mathematics and prior to its proof it was in the *Guinness Book of World Records* for "most difficult mathematical problems".

10.1.2 Four color theorem

Main article: Four color theorem

In mathematics, the four color theorem, or the four color map theorem, states that, given any separation of a plane into

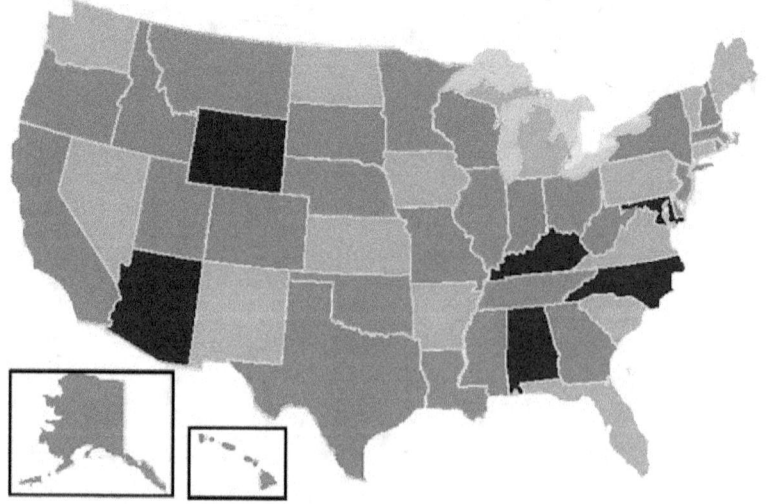

A four-coloring of a map of the states of the United States (ignoring lakes).

contiguous regions, producing a figure called a *map*, no more than four colors are required to color the regions of the map so that no two adjacent regions have the same color. Two regions are called *adjacent* if they share a common boundary that is not a corner, where corners are the points shared by three or more regions.[4] For example, in the map of the United States of America, Utah and Arizona are adjacent, but Utah and New Mexico, which only share a point that also belongs to Arizona and Colorado, are not.

Möbius mentioned the problem in his lectures as early as 1840.[5] The conjecture was first proposed on October 23, 1852 [6] when Francis Guthrie, while trying to color the map of counties of England, noticed that only four different colors were needed. The five color theorem, which has a short elementary proof, states that five colors suffice to color a map and was proven in the late 19th century (Heawood 1890); however, proving that four colors suffice turned out to be significantly harder. A number of false proofs and false counterexamples have appeared since the first statement of the four color theorem in 1852.

The four color theorem was proven in 1976 by Kenneth Appel and Wolfgang Haken. It was the first major theorem to be proved using a computer. Appel and Haken's approach started by showing that there is a particular set of 1,936 maps, each of which cannot be part of a smallest-sized counterexample to the four color theorem. (If they did appear, you could make a smaller counter-example.) Appel and Haken used a special-purpose computer program to confirm that each of these maps had this property. Additionally, any map that could potentially be a counterexample must have a portion that looks like one of these 1,936 maps. Showing this required hundreds of pages of hand analysis. Appel and Haken concluded that no smallest counterexamples exists because any must contain, yet do not contain, one of these 1,936 maps. This contradiction means there are no counterexamples at all and that the theorem is therefore true. Initially, their proof was not accepted by all mathematicians because the computer-assisted proof was infeasible for a human to check by hand (Swart 1980). Since then the proof has gained wider acceptance, although doubts remain (Wilson 2002, 216–222).

10.1.3   Hauptvermutung

Main article: Hauptvermutung

The Hauptvermutung (German for main conjecture) of geometric topology is the conjecture that any two triangulations of a triangulable space have a common refinement, a single triangulation that is a subdivision of both of them. It was originally formulated in 1908, by Steinitz and Tietze.

This conjecture is now known to be false. The non-manifold version was disproved by John Milnor[7] in 1961 using Reidemeister torsion.

The manifold version is true in dimensions $m \leq 3$. The cases $m = 2$ and 3 were proved by Tibor Radó and Edwin E. Moise[8] in the 1920s and 1950s, respectively.

10.1.4   Weil conjectures

Main article: Weil conjectures

In mathematics, the Weil conjectures were some highly influential proposals by André Weil (1949) on the generating functions (known as local zeta-functions) derived from counting the number of points on algebraic varieties over finite fields.

A variety V over a finite field with q elements has a finite number of rational points, as well as points over every finite field with q^k elements containing that field. The generating function has coefficients derived from the numbers Nk of points over the (essentially unique) field with q^k elements.

Weil conjectured that such *zeta-functions* should be rational functions, should satisfy a form of functional equation, and should have their zeroes in restricted places. The last two parts were quite consciously modeled on the Riemann zeta function and Riemann hypothesis. The rationality was proved by Dwork (1960), the functional equation by Grothendieck (1965), and the analogue of the Riemann hypothesis was proved by Deligne (1974)

10.1.5   Poincaré theorem

Main article: Poincaré theorem

In mathematics, the Poincaré conjecture is a theorem about the characterization of the 3-sphere, which is the hypersphere that bounds the unit ball in four-dimensional space. The conjecture states:

> Every simply connected, closed 3-manifold is homeomorphic to the 3-sphere.

An equivalent form of the conjecture involves a coarser form of equivalence than homeomorphism called homotopy equivalence: if a 3-manifold is *homotopy equivalent* to the 3-sphere, then it is necessarily *homeomorphic* to it.

Originally conjectured by Henri Poincaré, the theorem concerns a space that locally looks like ordinary three-dimensional space but is connected, finite in size, and lacks any boundary (a closed 3-manifold). The Poincaré conjecture claims that if such a space has the additional property that each loop in the space can be continuously tightened to a point, then it is necessarily a three-dimensional sphere. An analogous result has been known in higher dimensions for some time.

After nearly a century of effort by mathematicians, Grigori Perelman presented a proof of the conjecture in three papers made available in 2002 and 2003 on arXiv. The proof followed on from the program of Richard Hamilton to use the Ricci flow to attempt to solve the problem. Hamilton later introduced a modification of the standard Ricci flow, called *Ricci flow with surgery* to systematically excise singular regions as they develop, in a controlled way, but was unable to prove this method "converged" in three dimensions.[9] Perelman completed this portion of the proof. Several teams of mathematicians have verified that Perelman's proof is correct.

The Poincaré conjecture, before being proven, was one of the most important open questions in topology.

10.1.6 Riemann hypothesis

Main article: Riemann hypothesis

In mathematics, the Riemann hypothesis, proposed by Bernhard Riemann (1859), is a conjecture that the non-trivial zeros of the Riemann zeta function all have real part 1/2. The name is also used for some closely related analogues, such as the Riemann hypothesis for curves over finite fields.

The Riemann hypothesis implies results about the distribution of prime numbers. Along with suitable generalizations, some mathematicians consider it the most important unresolved problem in pure mathematics (Bombieri 2000). The Riemann hypothesis, along with the Goldbach conjecture, is part of Hilbert's eighth problem in David Hilbert's list of 23 unsolved problems; it is also one of the Clay Mathematics Institute Millennium Prize Problems.

10.1.7 P versus NP problem

Main article: P versus NP problem

The P versus NP problem is a major unsolved problem in computer science. Informally, it asks whether every problem whose solution can be quickly verified by a computer can also be quickly solved by a computer; it is widely conjectured that the answer is no. It was essentially first mentioned in a 1956 letter written by Kurt Gödel to John von Neumann. Gödel asked whether a certain NP complete problem could be solved in quadratic or linear time.[10] The precise statement of the P=NP problem was introduced in 1971 by Stephen Cook in his seminal paper "The complexity of theorem proving procedures"[11] and is considered by many to be the most important open problem in the field.[12] It is one of the seven Millennium Prize Problems selected by the Clay Mathematics Institute to carry a US$1,000,000 prize for the first correct solution.

10.1.8 Other conjectures

- Goldbach's conjecture
- The twin prime conjecture

- The Collatz conjecture

- The Manin conjecture

- The Maldacena conjecture

- The Langlands program[13] is a far-reaching web of these ideas of 'unifying conjectures' that link different subfields of mathematics, e.g. number theory and representation theory of Lie groups; some of these conjectures have since been proved.

10.2   Resolution of conjectures

10.2.1   Proof

Formal mathematics is based on *provable* truth. In mathematics, any number of cases supporting a conjecture, no matter how large, is insufficient for establishing the conjecture's veracity, since a single counterexample would immediately bring down the conjecture. Mathematical journals sometimes publish the minor results of research teams having extended the search for a counterexample farther than previously done. For instance, the Collatz conjecture, which concerns whether or not certain sequences of integers terminate, has been tested for all integers up to 1.2×10^{12} (over a trillion). However, the failure to find a counterexample after extensive search does not constitute a proof that no counterexample exists nor that the theorem is true, because the theorem might be false but with a very large minimal counterexample.

Instead, a theorem is considered proven only when it has been shown that it is logically impossible for it to be false. There are various methods of doing so; see Mathematical proof#Methods for details.

One method of proof, usable when there are only a finite number of cases that could lead to counterexamples, is known as "brute force": in this approach, all possible cases are considered and shown not to give counterexamples. Sometimes the number of cases is quite large, in which situation a brute-force proof may require as a practical matter the use of a computer algorithm to check all the cases: the validity of the 1976 and 1997 brute-force proofs of the four color theorem by computer was initially doubted, but was eventually confirmed in 2005 by theorem-proving software.

When a conjecture has been proven, it is no longer a conjecture but a theorem. Many important theorems were once conjectures, such as the Geometrization theorem (which resolved the Poincaré conjecture), Fermat's Last Theorem, and others.

10.2.2   Disproof

Conjectures disproven through counterexample are sometimes referred to as *false conjectures* (cf. the Pólya conjecture and Euler's sum of powers conjecture). In the case of the latter, the first counterexample found involved numbers in the millions, although subsequently it has been found that the minimal counterexample is smaller than that.

10.2.3   Undecidable conjectures

Not every conjecture ends up being proven true or false. The continuum hypothesis, which tries to ascertain the relative cardinality of certain infinite sets, was eventually shown to be undecidable (or independent) from the generally accepted set of axioms of set theory. It is therefore possible to adopt this statement, or its negation, as a new axiom in a consistent manner (much as we can take Euclid's parallel postulate as either true or false).

In this case, if a proof uses this statement, researchers will often look for a new proof that *doesn't* require the hypothesis (in the same way that it is desirable that statements in Euclidean geometry be proved using only the axioms of neutral geometry, i.e. no parallel postulate.) The one major exception to this in practice is the axiom of choice—unless studying this axiom in particular, the majority of researchers do not usually worry whether a result requires the axiom of choice.

10.3 Conditional proofs

Sometimes a conjecture is called a *hypothesis* when it is used frequently and repeatedly as an assumption in proofs of other results. For example, the Riemann hypothesis is a conjecture from number theory that (amongst other things) makes predictions about the distribution of prime numbers. Few number theorists doubt that the Riemann hypothesis is true. In anticipation of its eventual proof, some have proceeded to develop further proofs which are contingent on the truth of this conjecture. These are called *conditional proofs*: the conjectures assumed appear in the hypotheses of the theorem, for the time being.

These "proofs", however, would fall apart if it turned out that the hypothesis was false, so there is considerable interest in verifying the truth or falsity of conjectures of this type.

10.4 In other sciences

Karl Popper pioneered the use of the term "conjecture" in scientific philosophy.[14] Conjecture is related to hypothesis, which in science refers to a testable conjecture.

10.5 See also

- Hypotheticals

- List of conjectures

10.6 References

[1] *Oxford Dictionary of English* (2010 ed.).

[2] Schwartz, JL (1995). *Shuttling between the particular and the general: reflections on the role of conjecture and hypothesis in the generation of knowledge in science and mathematics*. p. 93.

[3] Ore, Oystein (1988) [1948], *Number Theory and Its History*, Dover, pp. 203–204, ISBN 978-0-486-65620-5

[4] Georges Gonthier (December 2008). "Formal Proof—The Four-Color Theorem". *Notices of the AMS* **55** (11): 1382–1393.From this paper: Definitions: A planar map is a set of pairwise disjoint subsets of the plane, called regions. A simple map is one whose regions are connected open sets. Two regions of a map are adjacent if their respective closures have a common point that is not a corner of the map. A point is a corner of a map if and only if it belongs to the closures of at least three regions. Theorem: The regions of any simple planar map can be colored with only four colors, in such a way that any two adjacent regions have different colors.

[5] W. W. Rouse Ball (1960) *The Four Color Theorem*, in Mathematical Recreations and Essays, Macmillan, New York, pp 222-232.

[6] Donald MacKenzie, *Mechanizing Proof: Computing, Risk, and Trust* (MIT Press, 2004) p103

[7] Milnor, John W. (1961). "Two complexes which are homeomorphic but combinatorially distinct". *Annals of Mathematics* **74** (2): 575–590. doi:10.2307/1970299. JSTOR 1970299. MR 133127.

[8] Moise, Edwin E. (1977). *Geometric Topology in Dimensions 2 and 3*. New York: New York : Springer-Verlag. ISBN 978-0-387-90220-3.

[9] Hamilton, Richard S. (1997). "Four-manifolds with positive isotropic curvature". *Communications in Analysis and Geometry* **5** (1): 1–92. MR 1456308. Zbl 0892.53018.

[10] Juris Hartmanis 1989, Gödel, von Neumann, and the P = NP problem, Bulletin of the European Association for Theoretical Computer Science, vol. 38, pp. 101–107

[11] Cook, Stephen (1971). "The complexity of theorem proving procedures". *Proceedings of the Third Annual ACM Symposium on Theory of Computing.* pp. 151–158.

[12] Lance Fortnow, *The status of the P versus NP problem*, Communications of the ACM 52 (2009), no. 9, pp. 78–86. doi:10.1145/1562164.1562186

[13] Langlands, Robert (1967), *Letter to Prof. Weil*

[14] Popper, Karl (2004). *Conjectures and refutations : the growth of scientific knowledge.* London: Routledge. ISBN 0-415-28594-1.

10.7 External links

- Open Problem Garden
- Unsolved Problems web site

Chapter 11

Logic

This article is about reasoning and its study. For other uses, see Logic (disambiguation).

Philosophical **Logic** (from the Ancient Greek: λογική, *logike*)[1] is the use and study of valid reasoning.[2][3] The study of logic features most prominently in the subjects of philosophy, mathematics, and computer science.

Logic was studied in several ancient civilizations, including India,[4] China,[5] Persia and Greece. In the West, logic was established as a formal discipline by Aristotle, who gave it a fundamental place in philosophy. The study of logic was part of the classical trivium, which also included grammar and rhetoric. Logic was further extended by Al-Farabi who categorized it into two separate groups (idea and proof). Later, Avicenna revived the study of logic and developed relationship between temporalis and the implication. In the East, logic was developed by Hindus, Buddhists and Jains.

Logic is often divided into three parts: inductive reasoning, abductive reasoning, and deductive reasoning.

11.1   The study of logic

The concept of logical form is central to logic, it being held that the validity of an argument is determined by its logical form, not by its content. Traditional Aristotelian syllogistic logic and modern symbolic logic are examples of formal logics.

- **Informal logic** is the study of natural language arguments. The study of fallacies is an especially important branch of informal logic. The dialogues of Plato[6] are good examples of informal logic.

- **Formal logic** is the study of inference with purely formal content. An inference possesses a *purely formal content* if it can be expressed as a particular application of a wholly abstract rule, that is, a rule that is not about any particular thing or property. The works of Aristotle contain the earliest known formal study of logic. Modern formal logic follows and expands on Aristotle.[7] In many definitions of logic, logical inference and inference with purely formal content are the same. This does not render the notion of informal logic vacuous, because no formal logic captures all of the nuances of natural language.

- **Symbolic logic** is the study of symbolic abstractions that capture the formal features of logical inference.[8][9] Symbolic logic is often divided into two branches: propositional logic and predicate logic.

- **Mathematical logic** is an extension of symbolic logic into other areas, in particular to the study of model theory, proof theory, set theory, and recursion theory.

11.1.1   Logical form

Main article: Logical form

Logic is generally considered **formal** when it analyzes and represents the *form* of any valid argument type. The form of an argument is displayed by representing its sentences in the formal grammar and symbolism of a logical language to make its content usable in formal inference. If one considers the notion of form too philosophically loaded, one could say that formalizing simply means translating English sentences into the language of logic.

This is called showing the *logical form* of the argument. It is necessary because indicative sentences of ordinary language show a considerable variety of form and complexity that makes their use in inference impractical. It requires, first, ignoring those grammatical features irrelevant to logic (such as gender and declension, if the argument is in Latin), replacing conjunctions irrelevant to logic (such as "but") with logical conjunctions like "and" and replacing ambiguous, or alternative logical expressions ("any", "every", etc.) with expressions of a standard type (such as "all", or the universal quantifier ∀).

Second, certain parts of the sentence must be replaced with schematic letters. Thus, for example, the expression "all As are Bs" shows the logical form common to the sentences "all men are mortals", "all cats are carnivores", "all Greeks are philosophers", and so on.

That the concept of form is fundamental to logic was already recognized in ancient times. Aristotle uses variable letters to represent valid inferences in *Prior Analytics*, leading Jan Łukasiewicz to say that the introduction of variables was "one of Aristotle's greatest inventions".[10] According to the followers of Aristotle (such as Ammonius), only the logical principles stated in schematic terms belong to logic, not those given in concrete terms. The concrete terms "man", "mortal", etc., are analogous to the substitution values of the schematic placeholders *A*, *B*, *C*, which were called the "matter" (Greek *hyle*) of the inference.

The fundamental difference between modern formal logic and traditional, or Aristotelian logic, lies in their differing analysis of the logical form of the sentences they treat.

- In the traditional view, the form of the sentence consists of (1) a subject (e.g., "man") plus a sign of quantity ("all" or "some" or "no"); (2) the copula, which is of the form "is" or "is not"; (3) a predicate (e.g., "mortal"). Thus: all men are mortal. The logical constants such as "all", "no" and so on, plus sentential connectives such as "and" and "or" were called "syncategorematic" terms (from the Greek *kategorei* – to predicate, and *syn* – together with). This is a fixed scheme, where each judgment has an identified quantity and copula, determining the logical form of the sentence.

- According to the modern view, the fundamental form of a simple sentence is given by a recursive schema, involving logical connectives, such as a quantifier with its bound variable, which are joined by juxtaposition to other sentences, which in turn may have logical structure.

- The modern view is more complex, since a single judgement of Aristotle's system involves two or more logical connectives. For example, the sentence "All men are mortal" involves, in term logic, two non-logical terms "is a man" (here *M*) and "is mortal" (here *D*): the sentence is given by the judgement A(M,D). In predicate logic, the sentence involves the same two non-logical concepts, here analyzed as $m(x)$ and $d(x)$, and the sentence is given by $\forall x.(m(x) \rightarrow d(x))$, involving the logical connectives for universal quantification and implication.

- But equally, the modern view is more powerful. Medieval logicians recognized the problem of multiple generality, where Aristotelian logic is unable to satisfactorily render such sentences as "Some guys have all the luck", because both quantities "all" and "some" may be relevant in an inference, but the fixed scheme that Aristotle used allows only one to govern the inference. Just as linguists recognize recursive structure in natural languages, it appears that logic needs recursive structure.

11.1.2   Deductive and inductive reasoning, and abductive inference

Deductive reasoning concerns what follows necessarily from given premises (if *a*, then *b*). However, inductive reasoning— the process of deriving a reliable inference from observations—is often included in the study of logic. Similarly, it is important to distinguish deductive validity and inductive validity (called "strength"). An inference is deductively valid if and only if there is no possible situation in which all the premises are true but the conclusion false. An inference is inductively strong if and only if its premises give some degree of probability to its conclusion.

The notion of deductive validity can be rigorously stated for systems of formal logic in terms of the well-understood notions of semantics. Inductive validity on the other hand requires us to define a reliable generalization of some set of observations. The task of providing this definition may be approached in various ways, some less formal than others; some of these definitions may use mathematical models of probability. For the most part this discussion of logic deals only with deductive logic.

Abduction[11] is a form of logical inference that goes from observation to a hypothesis that accounts for the reliable data (observation) and seeks to explain relevant evidence. The American philosopher Charles Sanders Peirce (1839–1914) first introduced the term as "guessing".[12] Peirce said that to *abduce* a hypothetical explanation a from an observed surprising circumstance b is to surmise that a may be true because then b would be a matter of course.[13] Thus, to abduce a from b involves determining that a is sufficient (or nearly sufficient), but not necessary, for b.

11.1.3 Consistency, validity, soundness, and completeness

Among the important properties that logical systems can have:

- **Consistency**, which means that no theorem of the system contradicts another.[14]

- **Validity**, which means that the system's rules of proof never allow a false inference from true premises. A logical system has the property of soundness when the logical system has the property of validity and uses only premises that prove true (or, in the case of axioms, are true by definition).[14]

- **Completeness**, of a logical system, which means that if a formula is true, it can be proven (if it is true, it is a theorem of the system).

- **Soundness**, the term soundness has multiple separate meanings, which creates a bit of confusion throughout the literature. Most commonly, soundness refers to logical systems, which means that if some formula can be proven in a system, then it is true in the relevant model/structure (if A is a theorem, it is true). This is the converse of completeness. A distinct, peripheral use of soundness refers to arguments, which means that the premises of a valid argument are true in the actual world.

Some logical systems do not have all four properties. As an example, Kurt Gödel's incompleteness theorems show that sufficiently complex formal systems of arithmetic cannot be consistent and complete;[9] however, first-order predicate logics not extended by specific axioms to be arithmetic formal systems with equality can be complete and consistent.[15]

11.1.4 Rival conceptions of logic

Main article: Definitions of logic

Logic arose (see below) from a concern with correctness of argumentation. Modern logicians usually wish to ensure that logic studies just those arguments that arise from appropriately general forms of inference. For example, Thomas Hofweber writes in the Stanford Encyclopedia of Philosophy that logic "does not, however, cover good reasoning as a whole. That is the job of the theory of rationality. Rather it deals with inferences whose validity can be traced back to the formal features of the representations that are involved in that inference, be they linguistic, mental, or other representations".[16]

By contrast, Immanuel Kant argued that logic should be conceived as the science of judgement, an idea taken up in Gottlob Frege's logical and philosophical work. But Frege's work is ambiguous in the sense that it is both concerned with the "laws of thought" as well as with the "laws of truth", i.e. it both treats logic in the context of a theory of the mind, and treats logic as the study of abstract formal structures.

11.2   History

Main article: History of logic

In Europe, logic was first developed by Aristotle.[17] Aristotelian logic became widely accepted in science and mathematics and remained in wide use in the West until the early 19th century.[18] Aristotle's system of logic was responsible for the introduction of hypothetical syllogism,[19] temporal modal logic,[20][21] and inductive logic,[22] as well as influential terms such as terms, predicables, syllogisms and propositions. In Europe during the later medieval period, major efforts were made to show that Aristotle's ideas were compatible with Christian faith. During the High Middle Ages, logic became a main focus of philosophers, who would engage in critical logical analyses of philosophical arguments, often using variations of the methodology of scholasticism. In 1323, William of Ockham's influential *Summa Logicae* was released. By the 18th century, the structured approach to arguments had degenerated and fallen out of favour, as depicted in Holberg's satirical play *Erasmus Montanus*.

The Chinese logical philosopher Gongsun Long (c. 325–250 BCE) proposed the paradox "One and one cannot become two, since neither becomes two."[23] In China, the tradition of scholarly investigation into logic, however, was repressed by the Qin dynasty following the legalist philosophy of Han Feizi.

In India, innovations in the scholastic school, called Nyaya, continued from ancient times into the early 18th century with the Navya-Nyaya school. By the 16th century, it developed theories resembling modern logic, such as Gottlob Frege's "distinction between sense and reference of proper names" and his "definition of number", as well as the theory of "restrictive conditions for universals" anticipating some of the developments in modern set theory.[24] Since 1824, Indian logic attracted the attention of many Western scholars, and has had an influence on important 19th-century logicians such as Charles Babbage, Augustus De Morgan, and George Boole.[25] In the 20th century, Western philosophers like Stanislaw Schayer and Klaus Glashoff have explored Indian logic more extensively.

The syllogistic logic developed by Aristotle predominated in the West until the mid-19th century, when interest in the foundations of mathematics stimulated the development of symbolic logic (now called mathematical logic). In 1854, George Boole published *An Investigation of the Laws of Thought on Which are Founded the Mathematical Theories of Logic and Probabilities*, introducing symbolic logic and the principles of what is now known as Boolean logic. In 1879, Gottlob Frege published *Begriffsschrift*, which inaugurated modern logic with the invention of quantifier notation. From 1910 to 1913, Alfred North Whitehead and Bertrand Russell published *Principia Mathematica*[8] on the foundations of mathematics, attempting to derive mathematical truths from axioms and inference rules in symbolic logic. In 1931, Gödel raised serious problems with the foundationalist program and logic ceased to focus on such issues.

The development of logic since Frege, Russell, and Wittgenstein had a profound influence on the practice of philosophy and the perceived nature of philosophical problems (see Analytic philosophy), and Philosophy of mathematics. Logic, especially sentential logic, is implemented in computer logic circuits and is fundamental to computer science. Logic is commonly taught by university philosophy departments, often as a compulsory discipline.

11.3   Types of logic

11.3.1   Syllogistic logic

Main article: Aristotelian logic

The *Organon* was Aristotle's body of work on logic, with the *Prior Analytics* constituting the first explicit work in formal logic, introducing the syllogistic.[26] The parts of syllogistic logic, also known by the name term logic, are the analysis of the judgements into propositions consisting of two terms that are related by one of a fixed number of relations, and the expression of inferences by means of syllogisms that consist of two propositions sharing a common term as premise, and a conclusion that is a proposition involving the two unrelated terms from the premises.

Aristotle's work was regarded in classical times and from medieval times in Europe and the Middle East as the very picture of a fully worked out system. However, it was not alone: the Stoics proposed a system of propositional logic that was studied by medieval logicians. Also, the problem of multiple generality was recognized in medieval times. Nonetheless, problems with syllogistic logic were not seen as being in need of revolutionary solutions.

Aristotle, 384–322 BCE.

Today, some academics claim that Aristotle's system is generally seen as having little more than historical value (though there is some current interest in extending term logics), regarded as made obsolete by the advent of propositional logic and the predicate calculus. Others use Aristotle in argumentation theory to help develop and critically question argumentation schemes that are used in artificial intelligence and legal arguments.

11.3.2 Propositional logic (sentential logic)

Main article: Propositional calculus

A propositional calculus or logic (also a sentential calculus) is a formal system in which formulae representing propositions can be formed by combining atomic propositions using logical connectives, and in which a system of formal proof rules establishes certain formulae as "theorems".

11.3.3 Predicate logic

Main article: Predicate logic

Predicate logic is the generic term for symbolic formal systems such as first-order logic, second-order logic, many-sorted logic, and infinitary logic.

Predicate logic provides an account of quantifiers general enough to express a wide set of arguments occurring in natural language. Aristotelian syllogistic logic specifies a small number of forms that the relevant part of the involved judgements may take. Predicate logic allows sentences to be analysed into subject and argument in several additional ways—allowing predicate logic to solve the problem of multiple generality that had perplexed medieval logicians.

The development of predicate logic is usually attributed to Gottlob Frege, who is also credited as one of the founders of analytical philosophy, but the formulation of predicate logic most often used today is the first-order logic presented in Principles of Mathematical Logic by David Hilbert and Wilhelm Ackermann in 1928. The analytical generality of predicate logic allowed the formalization of mathematics, drove the investigation of set theory, and allowed the development of Alfred Tarski's approach to model theory. It provides the foundation of modern mathematical logic.

Frege's original system of predicate logic was second-order, rather than first-order. Second-order logic is most prominently defended (against the criticism of Willard Van Orman Quine and others) by George Boolos and Stewart Shapiro.

11.3.4 Modal logic

Main article: Modal logic

In languages, modality deals with the phenomenon that sub-parts of a sentence may have their semantics modified by special verbs or modal particles. For example, "*We go to the games*" can be modified to give "*We should go to the games*", and "*We can go to the games*" and perhaps "*We will go to the games*". More abstractly, we might say that modality affects the circumstances in which we take an assertion to be satisfied.

Aristotle's logic is in large parts concerned with the theory of non-modalized logic. Although, there are passages in his work, such as the famous sea-battle argument in *De Interpretatione* § 9, that are now seen as anticipations of modal logic and its connection with potentiality and time, the earliest formal system of modal logic was developed by Avicenna, whom ultimately developed a theory of "temporally modalized" syllogistic.[27]

While the study of necessity and possibility remained important to philosophers, little logical innovation happened until the landmark investigations of Clarence Irving Lewis in 1918, who formulated a family of rival axiomatizations of the alethic modalities. His work unleashed a torrent of new work on the topic, expanding the kinds of modality treated to include deontic logic and epistemic logic. The seminal work of Arthur Prior applied the same formal language to treat temporal logic and paved the way for the marriage of the two subjects. Saul Kripke discovered (contemporaneously with rivals) his theory of frame semantics, which revolutionized the formal technology available to modal logicians and gave a new graph-theoretic way of looking at modality that has driven many applications in computational linguistics and computer science, such as dynamic logic.

11.3.5 Informal reasoning

Main article: Informal logic

The motivation for the study of logic in ancient times was clear: it is so that one may learn to distinguish good from bad arguments, and so become more effective in argument and oratory, and perhaps also to become a better person. Half of the works of Aristotle's Organon treat inference as it occurs in an informal setting, side by side with the development of the syllogistic, and in the Aristotelian school, these informal works on logic were seen as complementary to Aristotle's treatment of rhetoric.

This ancient motivation is still alive, although it no longer takes centre stage in the picture of logic; typically dialectical logic forms the heart of a course in critical thinking, a compulsory course at many universities.

Argumentation theory is the study and research of informal logic, fallacies, and critical questions as they relate to every day and practical situations. Specific types of dialogue can be analyzed and questioned to reveal premises, conclusions, and fallacies. Argumentation theory is now applied in artificial intelligence and law.

11.3.6 Mathematical logic

Main article: Mathematical logic

Mathematical logic really refers to two distinct areas of research: the first is the application of the techniques of formal logic to mathematics and mathematical reasoning, and the second, in the other direction, the application of mathematical techniques to the representation and analysis of formal logic.[28]

The earliest use of mathematics and geometry in relation to logic and philosophy goes back to the ancient Greeks such as Euclid, Plato, and Aristotle.[29] Many other ancient and medieval philosophers applied mathematical ideas and methods to their philosophical claims.[30]

One of the boldest attempts to apply logic to mathematics was undoubtedly the logicism pioneered by philosopher-logicians such as Gottlob Frege and Bertrand Russell: the idea was that mathematical theories were logical tautologies, and the programme was to show this by means to a reduction of mathematics to logic.[8] The various attempts to carry this out met with a series of failures, from the crippling of Frege's project in his *Grundgesetze* by Russell's paradox, to the defeat of Hilbert's program by Gödel's incompleteness theorems.

Both the statement of Hilbert's program and its refutation by Gödel depended upon their work establishing the second area of mathematical logic, the application of mathematics to logic in the form of proof theory.[31] Despite the negative nature of the incompleteness theorems, Gödel's completeness theorem, a result in model theory and another application of mathematics to logic, can be understood as showing how close logicism came to being true: every rigorously defined mathematical theory can be exactly captured by a first-order logical theory; Frege's proof calculus is enough to *describe* the whole of mathematics, though not *equivalent* to it.

If proof theory and model theory have been the foundation of mathematical logic, they have been but two of the four pillars of the subject. Set theory originated in the study of the infinite by Georg Cantor, and it has been the source of many of the most challenging and important issues in mathematical logic, from Cantor's theorem, through the status of the Axiom of Choice and the question of the independence of the continuum hypothesis, to the modern debate on large cardinal axioms.

Recursion theory captures the idea of computation in logical and arithmetic terms; its most classical achievements are the undecidability of the Entscheidungsproblem by Alan Turing, and his presentation of the Church–Turing thesis.[32] Today recursion theory is mostly concerned with the more refined problem of complexity classes—when is a problem efficiently solvable?—and the classification of degrees of unsolvability.[33]

11.3.7   Philosophical logic

Main article: Philosophical logic

Philosophical logic deals with formal descriptions of ordinary, non-specialist ("natural") language. Most philosophers assume that the bulk of everyday reasoning can be captured in logic if a method or methods to translate ordinary language into that logic can be found. Philosophical logic is essentially a continuation of the traditional discipline called "logic" before the invention of mathematical logic. Philosophical logic has a much greater concern with the connection between natural language and logic. As a result, philosophical logicians have contributed a great deal to the development of non-standard logics (e.g. free logics, tense logics) as well as various extensions of classical logic (e.g. modal logics) and non-standard semantics for such logics (e.g. Kripke's supervaluationism in the semantics of logic).

Logic and the philosophy of language are closely related. Philosophy of language has to do with the study of how our language engages and interacts with our thinking. Logic has an immediate impact on other areas of study. Studying logic and the relationship between logic and ordinary speech can help a person better structure his own arguments and critique the arguments of others. Many popular arguments are filled with errors because so many people are untrained in logic and unaware of how to formulate an argument correctly.[34][35]

11.3.8   Computational logic

Main articles: Computational logic and Logic in computer science

Logic cut to the heart of computer science as it emerged as a discipline: Alan Turing's work on the *Entscheidungsproblem* followed from Kurt Gödel's work on the incompleteness theorems. The notion of the general purpose computer that came from this work was of fundamental importance to the designers of the computer machinery in the 1940s.

In the 1950s and 1960s, researchers predicted that when human knowledge could be expressed using logic with mathematical notation, it would be possible to create a machine that reasons, or artificial intelligence. This was more difficult than expected because of the complexity of human reasoning. In logic programming, a program consists of a set of axioms and rules. Logic programming systems such as Prolog compute the consequences of the axioms and rules in order to answer a query.

Today, logic is extensively applied in the fields of Artificial Intelligence, and Computer Science, and these fields provide a rich source of problems in formal and informal logic. Argumentation theory is one good example of how logic is being applied to artificial intelligence. The ACM Computing Classification System in particular regards:

- Section F.3 on Logics and meanings of programs and F.4 on Mathematical logic and formal languages as part of the theory of computer science: this work covers formal semantics of programming languages, as well as work of formal methods such as Hoare logic;

- Boolean logic as fundamental to computer hardware: particularly, the system's section B.2 on Arithmetic and logic structures, relating to operatives AND, NOT, and OR;

- Many fundamental logical formalisms are essential to section I.2 on artificial intelligence, for example modal logic and default logic in Knowledge representation formalisms and methods, Horn clauses in logic programming, and description logic.

Furthermore, computers can be used as tools for logicians. For example, in symbolic logic and mathematical logic, proofs by humans can be computer-assisted. Using automated theorem proving the machines can find and check proofs, as well as work with proofs too lengthy to write out by hand.

11.3.9   Bivalence and the law of the excluded middle; non-classical logics

Main article: Principle of bivalence

The logics discussed above are all "bivalent" or "two-valued"; that is, they are most naturally understood as dividing propositions into true and false propositions. Non-classical logics are those systems that reject bivalence.

Hegel developed his own dialectic logic that extended Kant's transcendental logic but also brought it back to ground by assuring us that "neither in heaven nor in earth, neither in the world of mind nor of nature, is there anywhere such an abstract 'either–or' as the understanding maintains. Whatever exists is concrete, with difference and opposition in itself".[36]

In 1910, Nicolai A. Vasiliev extended the law of excluded middle and the law of contradiction and proposed the law of excluded fourth and logic tolerant to contradiction.[37] In the early 20th century Jan Łukasiewicz investigated the extension of the traditional true/false values to include a third value, "possible", so inventing ternary logic, the first multi-valued logic.[38]

Logics such as fuzzy logic have since been devised with an infinite number of "degrees of truth", represented by a real number between 0 and 1.[39]

Intuitionistic logic was proposed by L.E.J. Brouwer as the correct logic for reasoning about mathematics, based upon his rejection of the law of the excluded middle as part of his intuitionism. Brouwer rejected formalization in mathematics, but his student Arend Heyting studied intuitionistic logic formally, as did Gerhard Gentzen. Intuitionistic logic is of great interest to computer scientists, as it is a constructive logic and can be applied for extracting verified programs from proofs.

Modal logic is not truth conditional, and so it has often been proposed as a non-classical logic. However, modal logic is normally formalized with the principle of the excluded middle, and its relational semantics is bivalent, so this inclusion is disputable.

11.3.10 "Is logic empirical?"

Main article: Is logic empirical?

What is the epistemological status of the laws of logic? What sort of argument is appropriate for criticizing purported principles of logic? In an influential paper entitled "Is logic empirical?"[40] Hilary Putnam, building on a suggestion of W. V. Quine, argued that in general the facts of propositional logic have a similar epistemological status as facts about the physical universe, for example as the laws of mechanics or of general relativity, and in particular that what physicists have learned about quantum mechanics provides a compelling case for abandoning certain familiar principles of classical logic: if we want to be realists about the physical phenomena described by quantum theory, then we should abandon the principle of distributivity, substituting for classical logic the quantum logic proposed by Garrett Birkhoff and John von Neumann.[41]

Another paper of the same name by Sir Michael Dummett argues that Putnam's desire for realism mandates the law of distributivity.[42] Distributivity of logic is essential for the realist's understanding of how propositions are true of the world in just the same way as he has argued the principle of bivalence is. In this way, the question, "Is logic empirical?" can be seen to lead naturally into the fundamental controversy in metaphysics on realism versus anti-realism.

11.3.11 Implication: strict or material?

Main article: Paradox of entailment

The notion of implication formalized in classical logic does not comfortably translate into natural language by means of "if ... then ...", due to a number of problems called the paradoxes of material implication.

The first class of paradoxes involves counterfactuals, such as *If the moon is made of green cheese, then 2+2=5*, which are puzzling because natural language does not support the principle of explosion. Eliminating this class of paradoxes was the reason for C. I. Lewis's formulation of strict implication, which eventually led to more radically revisionist logics such as relevance logic.

The second class of paradoxes involves redundant premises, falsely suggesting that we know the succedent because of the antecedent: thus "if that man gets elected, granny will die" is materially true since granny is mortal, regardless of the man's election prospects. Such sentences violate the Gricean maxim of relevance, and can be modelled by logics that reject the principle of monotonicity of entailment, such as relevance logic.

11.3.12   Tolerating the impossible

Main article: Paraconsistent logic

Hegel was deeply critical of any simplified notion of the Law of Non-Contradiction. It was based on Leibniz's idea that this law of logic also requires a sufficient ground to specify from what point of view (or time) one says that something cannot contradict itself. A building, for example, both moves and does not move; the ground for the first is our solar system and for the second the earth. In Hegelian dialectic, the law of non-contradiction, of identity, itself relies upon difference and so is not independently assertable.

Closely related to questions arising from the paradoxes of implication comes the suggestion that logic ought to tolerate inconsistency. Relevance logic and paraconsistent logic are the most important approaches here, though the concerns are different: a key consequence of classical logic and some of its rivals, such as intuitionistic logic, is that they respect the principle of explosion, which means that the logic collapses if it is capable of deriving a contradiction. Graham Priest, the main proponent of dialetheism, has argued for paraconsistency on the grounds that there are in fact, true contradictions.[43]

11.3.13   Rejection of logical truth

The philosophical vein of various kinds of skepticism contains many kinds of doubt and rejection of the various bases on which logic rests, such as the idea of logical form, correct inference, or meaning, typically leading to the conclusion that there are no logical truths. Observe that this is opposite to the usual views in philosophical skepticism, where logic directs skeptical enquiry to doubt received wisdoms, as in the work of Sextus Empiricus.

Friedrich Nietzsche provides a strong example of the rejection of the usual basis of logic: his radical rejection of idealization led him to reject truth as a "... mobile army of metaphors, metonyms, and anthropomorphisms—in short ... metaphors which are worn out and without sensuous power; coins which have lost their pictures and now matter only as metal, no longer as coins."[44] His rejection of truth did not lead him to reject the idea of either inference or logic completely, but rather suggested that "logic [came] into existence in man's head [out] of illogic, whose realm originally must have been immense. Innumerable beings who made inferences in a way different from ours perished".[45] Thus there is the idea that logical inference has a use as a tool for human survival, but that its existence does not support the existence of truth, nor does it have a reality beyond the instrumental: "Logic, too, also rests on assumptions that do not correspond to anything in the real world".[46]

This position held by Nietzsche however, has come under extreme scrutiny for several reasons. He fails to demonstrate the validity of his claims and merely asserts them rhetorically. Although, since he is criticising the established criteria of validity, this does not undermine his position for one could argue that the demonstration of validity provided in the name of logic was just as rhetorically based. Some philosophers, such as Jürgen Habermas, claim his position is self-refuting—and accuse Nietzsche of not even having a coherent perspective, let alone a theory of knowledge.[47] Again, it is unclear if this is a decisive critique for the criteria of coherency and consistent theory are exactly what is under question. Georg Lukács, in his book *The Destruction of Reason*, asserts that, "Were we to study Nietzsche's statements in this area from a logico-philosophical angle, we would be confronted by a dizzy chaos of the most lurid assertions, arbitrary and violently incompatible."[48] Still, in this respect his "theory" would be a much better depicition of a confused and chaotic reality than any consistent and compatible theory. Bertrand Russell described Nietzsche's irrational claims with "He is fond of expressing himself paradoxically and with a view to shocking conventional readers" in his book *A History of Western Philosophy*.[49]

11.4   See also

- Digital electronics (also known as *digital logic* or logic gates)

- Fallacies

- List of logicians

- List of logic journals

- Logic puzzle

- Logic symbols

- Mathematics

  - List of mathematics articles
  - Outline of mathematics

- Metalogic

- Outline of logic

- Philosophy

  - List of philosophy topics
  - Outline of philosophy

- Reason

- *Straight and Crooked Thinking* (book)

- Table of logic symbols

- Truth

- Vector logic

11.5   Notes and references

[1] "possessed of reason, intellectual, dialectical, argumentative", also related to λόγος (*logos*), "word, thought, idea, argument, account, reason, or principle" (Liddell & Scott 1999; Online Etymology Dictionary 2001).

[2] Richard Henry Popkin; Avrum Stroll (1 July 1993). *Philosophy Made Simple*. Random House Digital, Inc. p. 238. ISBN 978-0-385-42533-9. Retrieved 5 March 2012.

[3] Jacquette, D. (2002). *A Companion to Philosophical Logic*. Wiley Online Library. p. 2.

[4] For example, Nyaya (syllogistic recursion) dates back 1900 years.

[5] Mohists and the school of Names date back at 2200 years.

[6] Plato (1976). Buchanan, Scott, ed. *The Portable Plato*. Penguin. ISBN 0-14-015040-4.

[7] Aristotle (2001). "Posterior Analytics". In Mckeon, Richard. *The Basic Works*. Modern Library. ISBN 0-375-75799-6.

[8] Whitehead, Alfred North; Russell, Bertrand (1967). *Principia Mathematica to *56*. Cambridge University Press. ISBN 0-521-62606-4.

[9] For a more modern treatment, see Hamilton, A. G. (1980). *Logic for Mathematicians*. Cambridge University Press. ISBN 0-521-29291-3.

[10] Łukasiewicz, Jan (1957). *Aristotle's syllogistic from the standpoint of modern formal logic* (2nd ed.). Oxford University Press. p. 7. ISBN 978-0-19-824144-7.

[11]   • Magnani, L. "Abduction, Reason, and Science: Processes of Discovery and Explanation". *Kluwer Academic Plenum Publishers, New York, 2001.* xvii. 205 pages. Hard cover, ISBN 0-306-46514-0.

          • R. Josephson, J. & G. Josephson, S. "Abductive Inference: Computation, Philosophy, Technology" *Cambridge University Press, New York & Cambridge (U.K.).* viii. 306 pages. Hard cover (1994), ISBN 0-521-43461-0, Paperback (1996), ISBN 0-521-57545-1.

          • Bunt, H. & Black, W. "Abduction, Belief and Context in Dialogue: Studies in Computational Pragmatics" *(Natural Language Processing, 1.) John Benjamins, Amsterdam & Philadelphia, 2000.* vi. 471 pages. Hard cover, ISBN 90-272-4983-0 (Europe),

          1-58619-794-2 (U.S.)

[12] Peirce, C. S.

          • "On the Logic of drawing History from Ancient Documents especially from Testimonies" (1901), *Collected Papers* v. 7, paragraph 219.

          • "PAP" ["Prolegomena to an Apology for Pragmatism"], MS 293 c. 1906, *New Elements of Mathematics* v. 4, pp. 319-320.

          • A Letter to F. A. Woods (1913), *Collected Papers* v. 8, paragraphs 385-388.

          (See under "Abduction" and "Retroduction" at *Commens Dictionary of Peirce's Terms.*)

[13] Peirce, C. S. (1903), Harvard lectures on pragmatism, *Collected Papers* v. 5, paragraphs 188–189.

[14] Bergmann, Merrie; Moor, James; Nelson, Jack (2009). *The Logic Book* (Fifth ed.). New York, NY: McGraw-Hill. ISBN 978-0-07-353563-0.

[15] Mendelson, Elliott (1964). "Quantification Theory: Completeness Theorems". *Introduction to Mathematical Logic.* Van Nostrand. ISBN 0-412-80830-7.

[16] Hofweber, T. (2004). "Logic and Ontology". In Zalta, Edward N. *Stanford Encyclopedia of Philosophy.*

[17] E.g., Kline (1972, p.53) wrote "A major achievement of Aristotle was the founding of the science of logic".

[18] "Aristotle", MTU Department of Chemistry.

[19] Jonathan Lear (1986). *"Aristotle and Logical Theory".* Cambridge University Press. p.34. ISBN 0-521-31178-0

[20] Simo Knuuttila (1981). *"Reforging the great chain of being: studies of the history of modal theories".* Springer Science & Business. p.71. ISBN 90-277-1125-9

[21] Michael Fisher, Dov M. Gabbay, Lluís Vila (2005). *"Handbook of temporal reasoning in artificial intelligence".* Elsevier. p.119. ISBN 0-444-51493-7

[22] Harold Joseph Berman (1983). *"Law and revolution: the formation of the Western legal tradition".* Harvard University Press. p.133. ISBN 0-674-51776-8

[23] The four Catuṣkoṭi logical divisions are formally very close to the four opposed propositions of the Greek tetralemma, which in turn are analogous to the four truth values of modern relevance logic Cf. Belnap (1977); Jayatilleke, K. N., (1967, The logic of four alternatives, in *Philosophy East and West*, University of Hawaii Press).

[24] Kisor Kumar Chakrabarti (June 1976). "Some Comparisons Between Frege's Logic and Navya-Nyaya Logic". *Philosophy and Phenomenological Research* (International Phenomenological Society) **36** (4): 554–563. doi:10.2307/2106873. JSTOR 2106873. This paper consists of three parts. The first part deals with Frege's distinction between sense and reference of proper names and a similar distinction in Navya-Nyaya logic. In the second part we have compared Frege's definition of number with the Navya-Nyaya definition of number. In the third part we have shown how the study of the so-called 'restrictive conditions for universals' in Navya-Nyaya logic anticipated some of the developments of modern set theory.

[25] Jonardon Ganeri (2001). *Indian logic: a reader.* Routledge. pp. vii, 5, 7. ISBN 0-7007-1306-9.

[26] "Aristotle". *Encyclopædia Britannica.*

[27] "History of logic: Arabic logic". Encyclopædia Britannica.

[28] Stolyar, Abram A. (1983). *Introduction to Elementary Mathematical Logic.* Dover Publications. p. 3. ISBN 0-486-64561-4.

[29] Barnes, Jonathan (1995). *The Cambridge Companion to Aristotle*. Cambridge University Press. p. 27. ISBN 0-521-42294-9.

[30] Aristotle (1989). *Prior Analytics*. Hackett Publishing Co. p. 115. ISBN 978-0-87220-064-7.

[31] Mendelson, Elliott (1964). "Formal Number Theory: Gödel's Incompleteness Theorem". *Introduction to Mathematical Logic*. Monterey, Calif.: Wadsworth & Brooks/Cole Advanced Books & Software. OCLC 13580200.

[32] Brookshear, J. Glenn (1989). "Computability: Foundations of Recursive Function Theory". *Theory of computation: formal languages, automata, and complexity*. Redwood City, Calif.: Benjamin/Cummings Pub. Co. ISBN 0-8053-0143-7.

[33] Brookshear, J. Glenn (1989). "Complexity". *Theory of computation: formal languages, automata, and complexity*. Redwood City, Calif.: Benjamin/Cummings Pub. Co. ISBN 0-8053-0143-7.

[34] Goldman, Alvin I. (1986), *Epistemology and Cognition*, Harvard University Press, p. 293, ISBN 9780674258969, untrained subjects are prone to commit various sorts of fallacies and mistakes.

[35] Demetriou, A.; Efklides, A., eds. (1994), *Intelligence, Mind, and Reasoning: Structure and Development*, Advances in Psychology **106**, Elsevier, p. 194, ISBN 9780080867601.

[36] Hegel, G. W. F (1971) [1817]. *Philosophy of Mind*. Encyclopedia of the Philosophical Sciences. trans. William Wallace. Oxford: Clarendon Press. p. 174. ISBN 0-19-875014-5.

[37] Joseph E. Brenner (3 August 2008). *Logic in Reality*. Springer. pp. 28–30. ISBN 978-1-4020-8374-7. Retrieved 9 April 2012.

[38] Zegarelli, Mark (2010), *Logic For Dummies*, John Wiley & Sons, p. 30, ISBN 9781118053072.

[39] Hájek, Petr (2006). "Fuzzy Logic". In Zalta, Edward N. *Stanford Encyclopedia of Philosophy*.

[40] Putnam, H. (1969). "Is Logic Empirical?". *Boston Studies in the Philosophy of Science* **5**.

[41] Birkhoff, G.; von Neumann, J. (1936). "The Logic of Quantum Mechanics". *Annals of Mathematics* (Annals of Mathematics) **37** (4): 823–843. doi:10.2307/1968621. JSTOR 1968621.

[42] Dummett, M. (1978). "Is Logic Empirical?". *Truth and Other Enigmas*. ISBN 0-674-91076-1.

[43] Priest, Graham (2008). "Dialetheism". In Zalta, Edward N. *Stanford Encyclopedia of Philosophy*.

[44] Nietzsche, 1873, On Truth and Lies in a Nonmoral Sense.

[45] Nietzsche, 1882, *The Gay Science*.

[46] Nietzsche, 1878, *Human, All Too Human*

[47] Babette Babich, Habermas, Nietzsche, and Critical Theory

[48] Georg Lukács. "The Destruction of Reason by Georg Lukács 1952". Marxists.org. Retrieved 2013-06-16.

[49] Russell, Bertrand (1945), *A History of Western Philosophy And Its Connection with Political and Social Circumstances from the Earliest Times to the Present Day* (PDF), Simon and Schuster, p. 762

11.6 Bibliography

- Nuel Belnap, (1977). "A useful four-valued logic". In Dunn & Eppstein, *Modern uses of multiple-valued logic*. Reidel: Boston.

- Józef Maria Bocheński (1959). *A précis of mathematical logic*. Translated from the French and German editions by Otto Bird. D. Reidel, Dordrecht, South Holland.

- Józef Maria Bocheński, (1970). *A history of formal logic*. 2nd Edition. Translated and edited from the German edition by Ivo Thomas. Chelsea Publishing, New York.

- Brookshear, J. Glenn (1989). *Theory of computation: formal languages, automata, and complexity*. Redwood City, Calif.: Benjamin/Cummings Pub. Co. ISBN 0-8053-0143-7.

- Cohen, R.S, and Wartofsky, M.W. (1974). *Logical and Epistemological Studies in Contemporary Physics*. Boston Studies in the Philosophy of Science. D. Reidel Publishing Company: Dordrecht, Netherlands. ISBN 90-277-0377-9.

- Finkelstein, D. (1969). "Matter, Space, and Logic". in R.S. Cohen and M.W. Wartofsky (eds. 1974).

- Gabbay, D.M., and Guenthner, F. (eds., 2001–2005). *Handbook of Philosophical Logic*. 13 vols., 2nd edition. Kluwer Publishers: Dordrecht.

- Hilbert, D., and Ackermann, W, (1928). *Grundzüge der theoretischen Logik* (*Principles of Mathematical Logic*). Springer-Verlag. OCLC 2085765

- Susan Haack, (1996). *Deviant Logic, Fuzzy Logic: Beyond the Formalism*, University of Chicago Press.

- Hodges, W., (2001). *Logic. An introduction to Elementary Logic*, Penguin Books.

- Hofweber, T., (2004), Logic and Ontology. *Stanford Encyclopedia of Philosophy*. Edward N. Zalta (ed.).

- Hughes, R.I.G., (1993, ed.). *A Philosophical Companion to First-Order Logic*. Hackett Publishing.

- Kline, Morris (1972). *Mathematical Thought From Ancient to Modern Times*. Oxford University Press. ISBN 0-19-506135-7.

- Kneale, William, and Kneale, Martha, (1962). *The Development of Logic*. Oxford University Press, London, UK.

- Liddell, Henry George; Scott, Robert. "Logikos". *A Greek-English Lexicon*. Perseus Project. Retrieved 8 May 2009.

- Mendelson, Elliott, (1964). *Introduction to Mathematical Logic*. Wadsworth & Brooks/Cole Advanced Books & Software: Monterey, Calif. OCLC 13580200

- Harper, Robert (2001). "Logic". *Online Etymology Dictionary*. Retrieved 8 May 2009.

- Smith, B., (1989). "Logic and the Sachverhalt". *The Monist* 72(1):52–69.

- Whitehead, Alfred North and Bertrand Russell, (1910). *Principia Mathematica*. Cambridge University Press: Cambridge, England. OCLC 1041146

11.7   External links

- Logic at PhilPapers

- Logic at the Indiana Philosophy Ontology Project

- Logic entry in the *Internet Encyclopedia of Philosophy*

- Hazewinkel, Michiel, ed. (2001), "Logical calculus", *Encyclopedia of Mathematics*, Springer, ISBN 978-1-55608-010-4

- An Outline for Verbal Logic

- Introductions and tutorials

  - An Introduction to Philosophical Logic, by Paul Newall, aimed at beginners.
  - forall x: an introduction to formal logic, by P.D. Magnus, covers sentential and quantified logic.
  - Logic Self-Taught: A Workbook (originally prepared for on-line logic instruction).
    - Nicholas Rescher. (1964). *Introduction to Logic*, St. Martin's Press.

- Essays

- "Symbolic Logic" and "The Game of Logic", Lewis Carroll, 1896.
- Math & Logic: The history of formal mathematical, logical, linguistic and methodological ideas. In *The Dictionary of the History of Ideas.*
- Online Tools
  - Interactive Syllogistic Machine A web based syllogistic machine for exploring fallacies, figures, terms, and modes of syllogisms.
- Reference material
  - Translation Tips, by Peter Suber, for translating from English into logical notation.
  - Ontology and History of Logic. An Introduction with an annotated bibliography.
- Reading lists
  - The London Philosophy Study Guide offers many suggestions on what to read, depending on the student's familiarity with the subject:
    - Logic & Metaphysics
    - Set Theory and Further Logic
    - Mathematical Logic

Chapter 12

Informal logic

Informal logic, intuitively, refers to the principles of logic and logical thought outside of a formal setting. However, perhaps because of the "informal" in the title, the precise definition of "informal logic" is a matter of some dispute.[1] Ralph H. Johnson and J. Anthony Blair define informal logic as "a branch of logic whose task is to develop non-formal standards, criteria, procedures for the analysis, interpretation, evaluation, criticism and construction of argumentation."[2] This definition reflects what had been implicit in their practice and what others[3][4][5] were doing in their informal logic texts.

Informal logic is associated with (informal) fallacies, critical thinking, the Thinking Skills Movement[6] and the interdisciplinary inquiry known as argumentation theory. Frans H. van Eemeren writes that the label "informal logic" covers a "collection of normative approaches to the study of reasoning in ordinary language that remain closer to the practice of argumentation than formal logic."[7]

12.1 History

Informal logic as a distinguished enterprise under this name emerged roughly in the late 1970s as a sub-field of philosophy. The naming of the field was preceded by the appearance of a number of textbooks that rejected the symbolic approach to logic on pedagogical grounds as inappropriate and unhelpful for introductory textbooks on logic for a general audience, for example Howard Kahane's *Logic and Contemporary Rhetoric*, subtitled "The Use of Reason in Everyday Life", first published in 1971. Kahane's textbook was described on the notice of his death in the *Proceedings And Addresses of the American Philosophical Association* (2002) as "a text in informal logic, [that] was intended to enable students to cope with the misleading rhetoric one frequently finds in the media and in political discourse. It was organized around a discussion of fallacies, and was meant to be a practical instrument for dealing with the problems of everyday life. [It has] ... gone through many editions; [it is] ... still in print; and the thousands upon thousands of students who have taken courses in which his text [was] ... used can thank Howard for contributing to their ability to dissect arguments and avoid the deceptions of deceitful rhetoric. He tried to put into practice the ideal of discourse that aims at truth rather than merely at persuasion. (Hausman et al. 2002)"[8][9] Other textbooks from the era taking this approach were Michael Scriven's *Reasoning* (Edgepress, 1976) and *Logical Self-Defense* by Ralph Johnson and J. Anthony Blair, first published in 1977.[8] Earlier precursors in this tradition can be considered Monroe Beardsley's *Practical Logic* (1950) and Stephen Toulmin's *The Uses of Argument* (1958).[10]

The field perhaps became recognized under its current name with the *First International Symposium on Informal Logic* held in 1978. Although initially motivated by a new pedagogical approach to undergraduate logic textbooks, the scope of the field was basically defined by a list of 13 problems and issues which Blair and Johnson included as an appendix to their keynote address at this symposium:[8][11]

- the theory of logical criticism
- the theory of argument

- the theory of fallacy

- the fallacy approach vs. the critical thinking approach

- the viability of the inductive/deductive dichotomy

- the ethics of argumentation and logical criticism

- the problem of assumptions and missing premises

- the problem of context

- methods of extracting arguments from context

- methods of displaying arguments

- the problem of pedagogy

- the nature, division and scope of informal logic

- the relationship of informal logic to other inquiries

David Hitchcock argues that the naming of the field was unfortunate, and that *philosophy of argument* would have been more appropriate. He argues that more undergraduate students in North America study informal logic than any other branch of philosophy, but that as of 2003 informal logic (or philosophy of argument) was not recognized as separate sub-field by the World Congress of Philosophy.[8] Frans H. van Eemeren wrote that "informal logic" is mainly an approach to argumentation advanced by a group of US and Canadian philosophers and largely based on the previous works of Stephen Toulmin and to a lesser extent those of Chaïm Perelman.[7]

Alongside the symposia, since 1983 the journal *Informal Logic* has been the publication of record of the field, with Blair and Johnson as initial editors, with the editorial board now including two other colleagues from the University of Windsor—Christopher Tindale and Hans V. Hansen.[12] Other journals that regularly publish articles on informal logic include *Argumentation* (founded in 1986), *Philosophy and Rhetoric, Argumentation and Advocacy* (the journal of the American Forensic Association), and *Inquiry: Critical Thinking Across the Disciplines* (founded in 1988).[13]

12.2 Proposed definitions

Johnson and Blair (2000) proposed the following definition: "Informal logic designates that branch of logic whose task is to develop non-formal[2] standards, criteria, procedures for the analysis, interpretation, evaluation, critique and construction of argumentation in everyday discourse." Their meaning of non-formal[2] is taken from Barth and Krabbe (1982), which is explained below.

To understand the definition above, one must understand "informal" which takes its meaning in contrast to its counterpart "formal." (This point was not made for a very long time, hence the nature of informal logic remained opaque, even to those involved in it, for a period of time.) Here it is helpful to have recourse[14] to Barth and Krabbe (1982:14f) where they distinguish three senses of the term "form." By "form[1]," Barth and Krabbe mean the sense of the term which derives from the Platonic idea of form—the ultimate metaphysical unit. Barth and Krabbe claim that most traditional logic is formal in this sense. That is, syllogistic logic is a logic of terms where the terms could naturally be understood as place-holders for Platonic (or Aristotelian) forms. In this first sense of "form," almost all logic is informal (not-formal). Understanding informal logic this way would be much too broad to be useful.

By "form[2]," Barth and Krabbe mean the form of sentences and statements as these are understood in modern systems of logic. Here validity is the focus: if the premises are true, the conclusion must then also be true. Now validity has to do with the logical form of the statement that makes up the argument. In this sense of "formal," most modern and contemporary logic is "formal." That is, such logics canonize the notion of logical form, and the notion of validity plays the central normative role. In this second sense of form, informal logic is not-formal, because it abandons the notion of logical form as the key to understanding the structure of arguments, and likewise retires validity as normative for the purposes of the evaluation of argument. It seems to many that validity is too stringent a requirement, that there are good arguments in which the conclusion is supported by the premises even though it does not follow necessarily from them (as validity requires). An argument in which the conclusion is thought to be "beyond reasonable doubt, given the premises" is sufficient in law to cause a person to be sentenced to death, even though it does not meet the standard of logical validity. This type of argument, based on accumulation of evidence rather than pure deduction, is called a conductive argument.

By "form[3]," Barth and Krabbe mean to refer to "procedures which are somehow regulated or regimented, which take place according to some set of rules." Barth and Krabbe say that "we do not defend formality[3] of all kinds and under all circumstances." Rather "we defend the thesis that verbal dialectics must have a certain form (i.e., must proceed according to certain rules) in order that one can speak of the discussion as being won or lost" (19). In this third sense of "form", informal logic can be formal, for there is nothing in the informal logic enterprise that stands opposed to the idea that argumentative discourse should be subject to norms, i.e., subject to rules, criteria, standards or procedures. Informal logic does present standards for the evaluation of argument, procedures for detecting missing premises etc.

Johnson and Blair (2000) noticed a limitation of their own definition, particularly with respect to "everyday discourse", which could indicate that it does not seek to understand specialized, domain-specific arguments made in natural languages. Consequently, they have argued that the crucial divide is between arguments made in formal languages and those made in natural languages.

Fisher and Scriven (1997) proposed a more encompassing definition, seeing informal logic as "the discipline which studies the practice of critical thinking and provides its intellectual spine". By "critical thinking" they understand "skilled and active interpretation and evaluation of observations and communications, information and argumentation."[15]

12.3 Criticisms

Some hold the view that informal logic is not a branch or subdiscipline of logic, or even the view that there cannot be such a thing as informal logic.[16][17][18] Massey criticizes informal logic on the grounds that it has no theory underpinning it. Informal logic, he says, requires detailed classification schemes to organize it, which in other disciplines is provided by the underlying theory. He maintains that there is no method of establishing the invalidity of an argument aside from the formal method, and that the study of fallacies may be of more interest to other disciplines, like psychology, than to philosophy and logic.[16]

12.4 Relation to critical thinking

See also: Critical thinking

Since the 1980s, informal logic has been partnered and even equated,[19] in the minds of many, with critical thinking. The precise definition of "critical thinking" is a subject of much dispute.[20] Critical thinking, as defined by Johnson, is the evaluation of an intellectual product (an argument, an explanation, a theory) in terms of its strengths and weaknesses.[20] While critical thinking will include evaluation of arguments and hence require skills of argumentation including informal logic, critical thinking requires additional abilities not supplied by informal logic, such as the ability to obtain and assess information and to clarify meaning. Also, many believe that critical thinking requires certain dispositions.[21] Understood in this way, "critical thinking" is a broad term for the attitudes and skills that are involved in analyzing and evaluating arguments. The critical thinking movement promotes critical thinking as an educational ideal. The movement emerged with great force in the '80s in North America as part of an ongoing critique of education as regards the thinking skills not being taught.

12.5 Relation to argumentation theory

See also: Argumentation theory

The social, communicative practice of argumentation can and should be distinguished from implication (or entailment)—a relationship between propositions; and from inference—a mental activity typically thought of as the drawing of a conclusion from premises. Informal logic may thus be said to be a logic of argumentation, as distinguished from implication and inference.[22]

Argumentation theory (or the theory of argumentation) has come to be the term that designates the theoretical study of argumentation. This study is interdisciplinary in the sense that no one discipline will be able to provide a complete account. A full appreciation of argumentation requires insights from logic (both formal and informal), rhetoric, communication theory, linguistics, psychology, and, increasingly, computer science. Since the 1970s, there has been significant agreement that there are three basic approaches to argumentation theory: the logical, the rhetorical and the dialectical. According to Wenzel,[23] the logical approach deals with the product, the dialectical with the process, and the rhetorical with the procedure. Thus, informal logic is one contributor to this inquiry, being most especially concerned with the norms of argument.

12.6   See also

- Argument map
- Informal fallacy
- Inference objection
- Lemma
- Philosophy of language
- Semantics

12.7   Footnotes

[1] See Johnson 1999 for a survey of definitions.

[2] Johnson, Ralph H., and Blair, J. Anthony (1987), "The Current State of Informal Logic", *Informal Logic*, 9(2–3), 147–151. Johnson & Blair added "... in everyday discourse" but in (2000), modified their definition, and broadened the focus now to include the sorts of argument that occurs not just in everyday discourse but also disciplined inquiry—what Weinstein (1990) calls "stylized discourse."

[3] Scriven, 1976

[4] Munson, 1976

[5] Fogelin, 1978

[6] Resnick, 1989

[7] Frans H. van Eemeren (2009). "The Study of Argumentation". In Andrea A. Lunsford, Kirt H. Wilson, Rosa A. Eberly. *The SAGE handbook of rhetorical studies*. SAGE. p. 117. ISBN 978-1-4129-0950-1.

[8] David Hitchcock, Informal logic 25 years later in *Informal Logic at 25: Proceedings of the Windsor Conference (OSSA 2003)*

[9] JSTOR 3218569

[10] Fisher (2004) p. vii

[11] J. Anthony Blair and Ralph H. Johnson (eds.), Informal Logic: The First International Symposium, 3-28. Pt. Reyes, CA: Edgepress

[12] http://ojs.uwindsor.ca/ojs/leddy/index.php/informal_logic/about/editorialTeam

[13]   Johnson and Blair (2000), p. 100

[14]   As Johnson (1999) does.

[15]   Johnson and Blair (2000), p. 95

[16]   Massey, 1981

[17]   Woods, 1980

[18]   Woods, 2000

[19]   Johnson (2000) takes the conflation to be part of the Network Problem and holds that settling the issue will require a theory of reasoning.

[20]   Johnson, 1992

[21]   Ennis, 1987

[22]   Johnson, 1999

[23]   Wenzel (1990)

12.8   References

- Barth, E. M., & Krabbe, E. C. W. (Eds.). (1982). From axiom to dialogue: A philosophical study of logics and argumentation. Berlin: Walter De Gruyter.

- Blair, J. A & Johnson, R.H. (1980). The recent development of informal logic. In J. Anthony Blair and Ralph H. Johnson (Eds.). Informal logic: The first international symposium, (pp. 3–28). Inverness, CA: Edgepress.

- Ennis, R.H. (1987). A taxonomy of critical thinking dispositions and abilities. In J.B. Baron and R.J. Sternberg (Eds.), Teaching critical thinking skills: Theory and practice, (pp. 9–26). New York: Freeman.

- Eemeren, F. H. van, & Grootendorst, R. (1992). Argumentation, communication and fallacies. Hillsdale, NJ: Lawrence Erlbaum Associates.

- Fisher, A. and Scriven, M. (1997). Critical thinking: Its definition and assessment. Point Reyes, CA: Edgepress

- Fisher, Alec (2004). *The logic of real arguments* (2nd ed.). Cambridge University Press. ISBN 978-0-521-65481-4.

- Govier, T. (1987). Problems in argument analysis and evaluation. Dordrecht: Foris.

- Govier, T. (1999). The Philosophy of Argument. Newport News, VA: Vale Press.

- Groarke, L. (2006). Informal Logic. Stanford Encyclopedia of Philosophy, from http://plato.stanford.edu/entries/logic-informal/

- Hitchcock, David (2007). "Informal logic and the concept of argument". In Jacquette, Dale. *Philosophy of logic*. Elsevier. ISBN 978-0-444-51541-4. preprint

- Johnson, R. H. (1992). The problem of defining critical thinking. In S. P. Norris (Ed.), The generalizability of critical thinking (pp. 38–53). New York: Teachers College Press. (Reprinted in Johnson (1996).)

- Johnson, R. H. (1996). The rise of informal logic. Newport News, VA: Vale Press

- Johnson, R. H. (1999). The relation between formal and informal logic. *Argumentation*, 13(3) 265-74.

- Johnson, R. H. (2000). Manifest rationality: A pragmatic theory of argument. Mahwah, NJ: Lawrence Erlbaum Associates.

- Johnson, R. H. & Blair, J. A. (1987). The current state of informal logic. *Informal Logic* 9, 147-51.

- Johnson, R. H. & Blair, J. A. (1996). Informal logic and critical thinking. In F. van Eemeren, R. Grootendorst, & F. Snoeck Henkemans (Eds.), Fundamentals of argumentation theory (pp. 383–86). Mahwah, NJ: Lawrence Erlbaum Associates

- Johnson, R. H. & Blair, J. A. (2002). Informal logic and the reconfiguration of logic. In D. Gabbay, R. H. Johnson, H.-J. Ohlbach and J. Woods (Eds.). Handbook of the logic of argument and inference: The turn towards the practical (pp. 339–396). Elsivier: North Holland.

- MacFarlane, J. (2005). Logical Constants. Stanford Encyclopedia of Philosophy.

- Massey, G. (1981). The fallacy behind fallacies. *Midwest Studies of Philosophy*, 6, 489-500.

- Munson, R. (1976). The way of words: an informal logic. Boston: Houghton Mifflin.

- Resnick, L. (1987). Education and learning to think. Washington, DC: National Academy Press..

- Walton, D. N. (1990). What is reasoning? What is an argument? *The Journal of Philosophy*, 87, 399-419.

- Weinstein, M. (1990) Towards a research agenda for informal logic and critical thinking. *Informal Logic*, 12, 121-143.

- Wenzel, J. 1990 Three perspectives on argumentation. In R Trapp and J Scheutz, (Eds.), Perspectives on argumentation: Essays in honour of Wayne Brockreide, 9-26 Waveland Press: Prospect Heights, IL

- Woods, J. (1980). What is informal logic? In J.A. Blair & R. H. Johnson (Eds.), Informal Logic: The First International Symposium (pp. 57–68). Point Reyes, CA: Edgepress.

12.8.1 Special journal issue

The open access issue 20(2) of *Informal Logic* from year 2000 groups a number of papers addressing foundational issues, based on the Panel on Informal Logic that was held at the 1998 World Congress of Philosophy, including:

- Hitchcock, D. (2000) The significance of informal logic for philosophy. *Informal Logic* 20(2), 129-138.

- Johnson, R. H. & Blair, J. A. (2000). Informal logic: An overview. *Informal Logic* 20(2): 93-99.

- Woods, J. (2000). How Philosophical is Informal Logic? *Informal Logic* 20(2): 139-167. 2000

12.8.2 Textbooks

- Kahane, H. (1971). Logic and contemporary rhetoric:The use of reasoning in everyday life. Belmont: Wadsworth. Still in print as Nancy Cavender; Howard Kahane (2009). *Logic and Contemporary Rhetoric: The Use of Reason in Everyday Life* (11th ed.). Cengage Learning. ISBN 978-0-495-80411-6.

- Scriven, M. (1976). Reasoning. New York. McGraw Hill.

- Johnson, R. H. & Blair, J. A. (1977). Logical self-defense. Toronto: McGraw-Hill Ryerson. US Edition. (2006). New York: Idebate Press.

- Fogelin, R.J. (1978). Understanding arguments: An introduction to informal logic. New York: Harcourt, Brace, Jovanovich. Still in print as Sinnott-Armstrong, Walter; Fogelin, Robert (2010), *Understanding Arguments: An Introduction to Informal Logic* (8th ed.), Belmont, California: Wadsworth Cengage Learning, ISBN 978-0-495-60395-5

- Stephen N. Thomas (1997). *Practical reasoning in natural language* (4th ed.). Prentice Hall. ISBN 978-0-13-678269-8.

- Irving M. Copi; Keith Burgess-Jackson (1996). *Informal logic* (3rd ed.). Prentice Hall. ISBN 978-0-13-229048-7.

- Woods, John, Andrew Irvine and Douglas Walton, 2004. Argument: Critical Thinking, Logic and the Fallacies. Toronto: Prentice Hall

- Groarke, Leo and Christopher Tindale, 2004. Good Reasoning Matters! (3rd edition). Toronto: Oxford University Press

- Douglas N. Walton (2008). *Informal logic: a pragmatic approach* (2nd ed.). Cambridge University Press. ISBN 978-0-521-71380-1.

- Trudy Govier (2009). *A Practical Study of Argument* (7th ed.). Cengage Learning. ISBN 978-0-495-60340-5.

12.9   External links

- Informal Logic entry by Leo Groarke in the *Stanford Encyclopedia of Philosophy*

Chapter 13

Deductive reasoning

Deductive reasoning, also **deductive logic** or **logical deduction** or, informally, **"top-down" logic**,[1] is the process of reasoning from one or more statements (premises) to reach a logically certain conclusion.[2] It differs from inductive reasoning or abductive reasoning.

Deductive reasoning links premises with conclusions. If all premises are true, the terms are clear, and the rules of deductive logic are followed, then the conclusion reached is necessarily true.

Deductive reasoning (top-down logic) contrasts with inductive reasoning (bottom-up logic) in the following way: In deductive reasoning, a conclusion is reached reductively by applying general rules that hold over the entirety of a closed domain of discourse, narrowing the range under consideration until only the conclusions is left. In inductive reasoning, the conclusion is reached by generalizing or extrapolating from, i.e., there is epistemic uncertainty. Note, however, that the inductive reasoning mentioned here is not the same as induction used in mathematical proofs – mathematical induction is actually a form of deductive reasoning.

13.1   Simple example

An example of a deductive argument:

1. All men are mortal.

2. Kass is a man.

3. Therefore, Kass is mortal.

The first premise states that all objects classified as "men" have the attribute "mortal". The second premise states that "Kass" is classified as a "man" – a member of the set "men". The conclusion then states that "Kass" must be "mortal" because he inherits this attribute from his classification as a "man".

13.2   Law of detachment

Main article: Modus ponens

The law of detachment (also known as **affirming the antecedent** and **Modus ponens**) is the first form of deductive reasoning. A single conditional statement is made, and a hypothesis (P) is stated. The conclusion (Q) is then deduced from the statement and the hypothesis. The most basic form is listed below:

1. $P \rightarrow Q$ (conditional statement)

2. P (hypothesis stated)

3. Q (conclusion deduced)

In deductive reasoning, we can conclude Q from P by using the law of detachment.[3] However, if the conclusion (Q) is given instead of the hypothesis (P) then there is no definitive conclusion.

The following is an example of an argument using the law of detachment in the form of an if-then statement:

1. If an angle satisfies $90° < A < 180°$, then A is an obtuse angle.

2. $A = 120°$.

3. A is an obtuse angle.

Since the measurement of angle A is greater than 90° and less than 180°, we can deduce that A is an obtuse angle. If however, we are given the conclusion that A is an obtuse angle we cannot deduce the premise that $A = 120°$.

13.3   Law of syllogism

The law of syllogism takes two conditional statements and forms a conclusion by combining the hypothesis of one statement with the conclusion of another. Here is the general form:

1. $P \rightarrow Q$

2. $Q \rightarrow R$

3. Therefore, $P \rightarrow R$.

The following is an example:

1. If Larry is sick, then he will be absent.

2. If Larry is absent, then he will miss his classwork.

3. Therefore, if Larry is sick, then he will miss his classwork.

We deduced the final statement by combining the hypothesis of the first statement with the conclusion of the second statement. We also allow that this could be a false statement. This is an example of the Transitive Property in mathematics. The Transitive Property is sometimes phrased in this form:

1. $A = B$.

2. $B = C$.

3. Therefore $A = C$.

13.4 Law of contrapositive

Main article: Modus tollens

The law of contrapositive states that, in a conditional, if the conclusion is false, then the hypothesis must be false also. The general form is the following:

1. $P \rightarrow Q$.

2. ~Q.

3. Therefore we can conclude ~P.

The following are examples:

1. If it is raining, then there are clouds in the sky.

2. There are no clouds in the sky.

3. Thus, it is not raining.

13.5 Validity and soundness

Deductive arguments are evaluated in terms of their *validity* and *soundness*.

An argument is valid if it is impossible for its premises to be true while its conclusion is false. In other words, the conclusion must be true if the premises are true. An argument can be valid even though the premises are false.

An argument is sound if it is valid and the premises are true.

It is possible to have a deductive argument that is logically valid but is not sound. Fallacious arguments often take that form.

The following is an example of an argument that is valid, but not sound:

1. Everyone who eats carrots is a quarterback.

2. John eats carrots.

3. Therefore, John is a quarterback.

The example's first premise is false – there are people who eat carrots and are not quarterbacks – but the conclusion must be true, so long as the premises are true (i.e. it is impossible for the premises to be true and the conclusion false). Therefore the argument is valid, but not sound. Generalizations are often used to make invalid arguments, such as "everyone who eats carrots is a quarterback." Not everyone who eats carrots is a quarterback, thus proving the flaw of such arguments.

In this example, the first statement uses categorical reasoning, saying that all carrot-eaters are definitely quarterbacks. This theory of deductive reasoning – also known as term logic – was developed by Aristotle, but was superseded by propositional (sentential) logic and predicate logic.

Deductive reasoning can be contrasted with inductive reasoning, in regards to validity and soundness. In cases of inductive reasoning, even though the premises are true and the argument is "valid", it is possible for the conclusion to be false (determined to be false with a counterexample or other means).

13.6 History

Aristotle started documenting deductive reasoning in the 4th century BC.[4]

13.7   Education

Deductive reasoning is generally thought of as a skill that develops without any formal teaching or training. As a result of this belief, deductive reasoning skills are not taught in secondary schools, where students are expected to use reasoning more often and at a higher level.[5] It is in high school, for example, that students have an abrupt introduction to mathematical proofs – which rely heavily on deductive reasoning.[5]

13.8   See also

- Argument (logic)

- Logic

- Mathematical logic

- Abductive reasoning

- Analogical reasoning

- Correspondence theory of truth

- Defeasible reasoning

- Decision making

- Decision theory

- Fallacy

- Fault Tree Analysis

- Geometry

- Hypothetico-deductive method

- Inquiry

- Mathematical induction

- Inductive reasoning

- Inference

- Logical consequence

- Natural deduction

- Propositional calculus

- Retroductive reasoning

- Scientific method

- Theory of justification

- Soundness

- Syllogism

13.9 References

[1] Deduction & Induction, Research Methods Knowledge Base

[2] Sternberg, R. J. (2009). *Cognitive Psychology*. Belmont, CA: Wadsworth. p. 578. ISBN 978-0-495-50629-4.

[3] Guide to Logic

[4] Evans, Jonathan St. B. T.; Newstead, Stephen E.; Byrne, Ruth M. J., eds. (1993). *Human Reasoning: The Psychology of Deduction* (Reprint ed.). Psychology Press. p. 4. ISBN 9780863773136. Retrieved 2015-01-26. In one sense [...] one can see the psychology of deductive reasoning as being as old as the study of logic, which originated in the writings of Aristotle.

[5] Stylianides, G. J.; Stylianides (2008). "A. J.". *Mathematical Thinking and Learning* **10** (2): 103–133. doi:10.1080/109860607018544.

13.10 Further reading

- Vincent F. Hendricks, *Thought 2 Talk: A Crash Course in Reflection and Expression*, New York: Automatic Press / VIP, 2005, ISBN 87-991013-7-8

- Philip Johnson-Laird, Ruth M. J. Byrne, *Deduction*, Psychology Press 1991, ISBN 978-0-86377-149-1

- Zarefsky, David, *Argumentation: The Study of Effective Reasoning Parts I and II*, The Teaching Company 2002

- Bullemore, Thomas, * The Pragmatic Problem of Induction.

13.11 External links

- Deductive reasoning at PhilPapers

- Deductive reasoning at the Indiana Philosophy Ontology Project

- Deductive reasoning entry in the *Internet Encyclopedia of Philosophy*

Chapter 14

Inductive reasoning

"Inductive inference" redirects here. For the technique in mathematical proof, see Mathematical induction.

Inductive reasoning (as opposed to *deductive* reasoning or *abductive* reasoning) is reasoning in which the premises seek to supply strong evidence for (not absolute proof of) the truth of the conclusion. While the conclusion of a deductive argument is certain, the truth of the conclusion of an inductive argument is *probable*, based upon the evidence given.[1]

The philosophical definition of inductive reasoning is more nuanced than simple progression from particular/individual instances to broader generalizations. Rather, the premises of an inductive logical argument indicate some degree of support (inductive probability) for the conclusion but do not entail it; that is, they suggest truth but do not ensure it. In this manner, there is the possibility of moving from general statements to individual instances (for example, statistical syllogisms, discussed below).

Many dictionaries define inductive reasoning as reasoning that derives general principles from specific observations, though some sources disagree with this usage.[2]

14.1 Description

Inductive reasoning is inherently uncertain. It only deals in degrees to which, given the premises, the conclusion is *credible* according to some theory of evidence. Examples include a many-valued logic, Dempster–Shafer theory, or probability theory with rules for inference such as Bayes' rule. Unlike deductive reasoning, it does not rely on universals holding over a closed domain of discourse to draw conclusions, so it can be applicable even in cases of epistemic uncertainty (technical issues with this may arise however; for example, the second axiom of probability is a closed-world assumption).[3]

An example of an inductive argument:

> 100% of biological life forms that we know of depend on liquid water to exist.

> Therefore, if we discover a new biological life form it will probably depend on liquid water to exist.

This argument could have been made every time a new biological life form was found, and would have been correct every time; however, it is still possible that in the future a biological life form not requiring water could be discovered.

As a result, the argument may be stated less formally as:

> All biological life forms that we know of depend on liquid water to exist.

> All biological life probably depends on liquid water to exist.

14.2 Inductive vs. deductive reasoning

Unlike deductive arguments, inductive reasoning allows for the possibility that the conclusion is false, even if all of the premises are true.[4] Instead of being valid or invalid, inductive arguments are either *strong* or *weak*, which describes how *probable* it is that the conclusion is true.[5]

Given that "if *A* is true then *B*, *C*, and *D* are true", an example of deduction would be "*A* is true therefore we can deduce that *B*, *C*, and *D* are true". An example of induction would be "*B*, *C*, and *D* are observed to be true therefore *A* may be true". *A* is a reasonable explanation for *B*, *C*, and *D* being true.

For example:

> A large enough asteroid impact would create a very large crater and cause a severe impact winter that could drive the non-avian dinosaurs to extinction.

> We observe that there is a very large crater in the gulf of Mexico dating to very near the time of the extinction of the non-avian dinosaurs

> Therefore it is possible that this impact could explain why the non-avian dinosaurs went extinct.

Note however that this is not necesarily the case. Other events also coincide with the extinction of the non-avian dinosaurs. For example the Deccan Traps in India.

A classical example of an *incorrect* inductive argument was presented by John Vickers:

> All of the swans we have seen are white.

> Therefore, all swans are white.

Note that this definition of *inductive* reasoning excludes mathematical induction, which is a form of *deductive* reasoning.

14.3 Criticism

Main article: Problem of induction

Inductive reasoning has been criticized by thinkers as diverse as Sextus Empiricus[6] and Karl Popper.[7]

The classic philosophical treatment of the problem of induction was given by the Scottish philosopher David Hume.[8]

Although the use of inductive reasoning demonstrates considerable success, its application has been questionable. Recognizing this, Hume highlighted the fact that our mind draws uncertain conclusions from relatively limited experiences. In deduction, the truth value of the conclusion is based on the truth of the premise. In induction, however, the dependence on the premise is always uncertain. As an example, let's assume "all ravens are black." The fact that there are numerous black ravens supports the assumption. However, the assumption becomes inconsistent with the fact that there are white ravens. Therefore, the general rule of "all ravens are black" is inconsistent with the existence of the white raven. Hume further argued that it is impossible to justify inductive reasoning: specifically, that it cannot be justified deductively, so our only option is to justify it inductively. Since this is circular he concluded that our use of induction is unjustifiable with the help of "Hume's Fork".[9]

However, Hume then stated that even if induction were proved unreliable, we would still have to rely on it. So instead of a position of severe skepticism, Hume advocated a practical skepticism based on common sense, where the inevitability of induction is accepted.[10]

14.3.1 Biases

Inductive reasoning is also known as hypothesis construction because any conclusions made are based on current knowledge and predictions. As with deductive arguments, biases can distort the proper application of inductive argument, thereby preventing the reasoner from forming the most logical conclusion based on the clues. Examples of these biases include the availability heuristic, confirmation bias, and the predictable-world bias.

The availability heuristic causes the reasoner to depend primarily upon information that is readily available to him/her. People have a tendency to rely on information that is easily accessible in the world around them. For example, in surveys, when people are asked to estimate the percentage of people who died from various causes, most respondents would choose the causes that have been most prevalent in the media such as terrorism, and murders, and airplane accidents rather than causes such as disease and traffic accidents, which have been technically "less accessible" to the individual since they are not emphasized as heavily in the world around him/her.

The confirmation bias is based on the natural tendency to confirm rather than to deny a current hypothesis. Research has demonstrated that people are inclined to seek solutions to problems that are more consistent with known hypotheses rather than attempt to refute those hypotheses. Often, in experiments, subjects will ask questions that seek answers that fit established hypotheses, thus confirming these hypotheses. For example, if it is hypothesized that Sally is a sociable individual, subjects will naturally seek to confirm the premise by asking questions that would produce answers confirming that Sally is in fact a sociable individual.

The predictable-world bias revolves around the inclination to perceive order where it has not been proved to exist, either at all or at a particular level of abstraction. Gambling, for example, is one of the most popular examples of predictable-world bias. Gamblers often begin to think that they see simple and obvious patterns in the outcomes and, therefore, believe that they are able to predict outcomes based upon what they have witnessed. In reality, however, the outcomes of these games are difficult to predict and highly complex in nature. However, in general, people tend to seek some type of simplistic order to explain or justify their beliefs and experiences, and it is often difficult for them to realise that their perceptions of order may be entirely different from the truth.[11]

14.4   Types

14.4.1   Generalization

A generalization (more accurately, an *inductive generalization*) proceeds from a premise about a sample to a conclusion about the population.

> The proportion Q of the sample has attribute A.
>
> Therefore:
>
> The proportion Q of the population has attribute A.

Example

There are 20 balls—either black or white—in an urn. To estimate their respective numbers, you draw a sample of four balls and find that three are black and one is white. A good inductive generalization would be that there are 15 black, and five white, balls in the urn.

How much the premises support the conclusion depends upon (a) the number in the sample group, (b) the number in the population, and (c) the degree to which the sample represents the population (which may be achieved by taking a random sample). The hasty generalization and the biased sample are generalization fallacies.

14.4.2   Statistical syllogism

Main article: Statistical syllogism

A statistical syllogism proceeds from a generalization to a conclusion about an individual.

A proportion Q of population P has attribute A.

An individual X is a member of P.

Therefore:

There is a probability which corresponds to Q that X has A.

The proportion in the first premise would be something like "3/5ths of", "all", "few", etc. Two dicto simpliciter fallacies can occur in statistical syllogisms: "accident" and "converse accident".

14.4.3 Simple induction

Simple induction proceeds from a premise about a sample group to a conclusion about another individual.

Proportion Q of the known instances of population P has attribute A.

Individual I is another member of P.

Therefore:

There is a probability corresponding to Q that I has A.

This is a combination of a generalization and a statistical syllogism, where the conclusion of the generalization is also the first premise of the statistical syllogism.

Argument from analogy

Main article: Argument from analogy

The process of analogical inference involves noting the shared properties of two or more things, and from this basis inferring that they also share some further property:[12]

P and Q are similar in respect to properties a, b, and c.

Object P has been observed to have further property x.

Therefore, Q probably has property x also.

Analogical reasoning is very frequent in common sense, science, philosophy and the humanities, but sometimes it is accepted only as an auxiliary method. A refined approach is case-based reasoning.[13]

14.4.4 Causal inference

A causal inference draws a conclusion about a causal connection based on the conditions of the occurrence of an effect. Premises about the correlation of two things can indicate a causal relationship between them, but additional factors must be confirmed to establish the exact form of the causal relationship.

14.4.5 Prediction

A prediction draws a conclusion about a future individual from a past sample.

Proportion Q of observed members of group G have had attribute A.

Therefore:

There is a probability corresponding to Q that other members of group G will have attribute A when next observed.

14.5   Bayesian inference

As a logic of induction rather than a theory of belief, Bayesian inference does not determine which beliefs are *a priori* rational, but rather determines how we should rationally change the beliefs we have when presented with evidence. We begin by committing to a prior probability for a hypothesis based on logic or previous experience, and when faced with evidence, we adjust the strength of our belief in that hypothesis in a precise manner using Bayesian logic.

14.6   Inductive inference

Around 1960, Ray Solomonoff founded the theory of universal inductive inference, the theory of prediction based on observations; for example, predicting the next symbol based upon a given series of symbols. This is a formal inductive framework that combines algorithmic information theory with the Bayesian framework. Universal inductive inference is based on solid philosophical foundations,[14] and can be considered as a mathematically formalized Occam's razor. Fundamental ingredients of the theory are the concepts of algorithmic probability and Kolmogorov complexity.

14.7   See also

- Abductive reasoning
- Algorithmic information theory
- Algorithmic probability
- Analogy
- Bayesian probability
- Counterinduction
- Deductive reasoning
- Explanation
- Failure mode and effects analysis
- Falsifiability
- Grammar induction
- Inductive inference
- Inductive logic programming
- Inductive probability
- Inductive programming
- Inductive reasoning aptitude
- Inquiry
- Kolmogorov complexity
- Lateral thinking
- Laurence Jonathan Cohen
- Logic

- Logical positivism

- Machine learning

- Mathematical induction

- Mill's Methods

- Minimum description length

- Minimum message length

- Open world assumption

- Raven paradox

- Recursive Bayesian estimation

- Retroduction

- Solomonoff's theory of inductive inference

- Statistical inference

- Stephen Toulmin

- Universal artificial intelligence

14.8 References

[1] Copi, I. M.; Cohen, C.; Flage, D. E. (2007). *Essentials of Logic* (Second ed.). Upper Saddle River, NJ: Pearson Education. ISBN 978-0-13-238034-8.

[2] "Deductive and Inductive Arguments", *Internet Encyclopedia of Philosophy*, Some dictionaries define "deduction" as reasoning from the general to specific and "induction" as reasoning from the specific to the general. While this usage is still sometimes found even in philosophical and mathematical contexts, for the most part, it is outdated.

[3] Kosko, Bart (1990). "Fuzziness vs. Probability". *International Journal of General Systems* **17** (1): 211–240. doi:10.1080/03081079008935108

[4] John Vickers. The Problem of Induction. The Stanford Encyclopedia of Philosophy.

[5] Herms, D. "Logical Basis of Hypothesis Testing in Scientific Research" (PDF).

[6] Sextus Empiricus, *Outlines Of Pyrrhonism*. Trans. R.G. Bury, Harvard University Press, Cambridge, Massachusetts, 1933, p. 283.

[7] Popper, Karl R.; Miller, David W. (1983). "A proof of the impossibility of inductive probability". *Nature* **302** (5910): 687–688. doi:10.1038/302687a0.

[8] David Hume (1910) [1748]. *An Enquiry concerning Human Understanding*. P.F. Collier & Son. ISBN 0-19-825060-6.

[9] Vickers, John. "The Problem of Induction" (Section 2). *Stanford Encyclopedia of Philosophy*. 21 June 2010

[10] Vickers, John. "The Problem of Induction" (Section 2.1). *Stanford Encyclopedia of Philosophy*. 21 June 2010.

[11] Gray, Peter (2011). *Psychology* (Sixth ed.). New York: Worth. ISBN 978-1-4292-1947-1.

[12] Baronett, Stan (2008). *Logic*. Upper Saddle River, NJ: Pearson Prentice Hall. pp. 321–325.

[13] For more information on inferences by analogy, see Juthe, 2005.

[14] Rathmanner, Samuel; Hutter, Marcus (2011). "A Philosophical Treatise of Universal Induction". *Entropy* **13** (6): 1076–1136. doi:10.3390/e13061076.

14.9   Further reading

- Cushan, Anna-Marie (1983/2014). *Investigation into Facts and Values: Groundwork for a theory of moral conflict resolution*. [Thesis, Melbourne University], Ondwelle Publications (online): Melbourne.

- Herms, D. "Logical Basis of Hypothesis Testing in Scientific Research" (PDF).

- Kemerling, G. (27 October 2001). "Causal Reasoning".

- Holland, J. H.; Holyoak, K. J.; Nisbett, R. E.; Thagard, P. R. (1989). *Induction: Processes of Inference, Learning, and Discovery*. Cambridge, MA, USA: MIT Press. ISBN 0-262-58096-9.

- Holyoak, K.; Morrison, R. (2005). *The Cambridge Handbook of Thinking and Reasoning*. New York: Cambridge University Press. ISBN 978-0-521-82417-0.

14.10   External links

- Confirmation and Induction entry in the *Internet Encyclopedia of Philosophy*

- Inductive Logic entry in the *Stanford Encyclopedia of Philosophy*

- Inductive reasoning at PhilPapers

- Inductive reasoning at the Indiana Philosophy Ontology Project

- *Four Varieties of Inductive Argument* from the Department of Philosophy, University of North Carolina at Greensboro.

- *Properties of Inductive Reasoning* PDF (166 KiB), a psychological review by Evan Heit of the University of California, Merced.

- *The Mind, Limber* An article which employs the film The Big Lebowski to explain the value of inductive reasoning.

- The Pragmatic Problem of Induction, by Thomas Bullemore

Chapter 15

Empirical evidence

"Empirical" redirects here. For other uses, see Empirical (disambiguation).

Empirical evidence, **data**, or **knowledge**, also known as **sense experience**, is a collective term for the knowledge or source of knowledge acquired by means of the senses, particularly by observation and experimentation.[1] The term comes from the Greek word for experience, ἐμπειρία (*empeiría*). After Kant, it is common in philosophy to call the knowledge thus gained *a posteriori* **knowledge**. This is contrasted with *a priori* knowledge, the knowledge accessible from pure reason alone.

15.1 Meaning

Empirical evidence is information that justifies a belief in the truth or falsity of a claim. In the empiricist view, one can claim to have knowledge only when one has a true belief based on empirical evidence. This stands in contrast to the rationalist view under which reason or reflection alone is considered evidence for the truth or falsity of some propositions.[2] The senses are the primary source of empirical evidence. Although other sources of evidence, such as memory and the testimony of others, ultimately trace back to some sensory experience, they are considered secondary, or indirect.[2]

In another sense, empirical evidence may be synonymous with the outcome of an experiment. In this sense, an empirical result is a unified confirmation. In this context, the term *semi-empirical* is used for qualifying theoretical methods that use, in part, basic axioms or postulated scientific laws and experimental results. Such methods are opposed to theoretical *ab initio* methods, which are purely deductive and based on first principles.

In science, empirical evidence is required for a hypothesis to gain acceptance in the scientific community. Normally, this validation is achieved by the scientific method of hypothesis commitment, experimental design, peer review, adversarial review, reproduction of results, conference presentation and journal publication. This requires rigorous communication of hypothesis (usually expressed in mathematics), experimental constraints and controls (expressed necessarily in terms of standard experimental apparatus), and a common understanding of measurement.

Statements and arguments depending on empirical evidence are often referred to as *a posteriori* ("following experience") as distinguished from *a priori* (preceding it). *A priori* knowledge or justification is independent of experience (for example "All bachelors are unmarried"), whereas *a posteriori* knowledge or justification is dependent on experience or empirical evidence (for example "Some bachelors are very happy"). The notion of the distinction between *a priori* and *a posteriori* as tantamount to the distinction between empirical and non-empirical knowledge comes from Kant's *Critique of Pure Reason*.[3]

The standard positivist view of empirically acquired information has been that observation, experience, and experiment serve as neutral arbiters between competing theories. However, since the 1960s, a persistent critique most associated with Thomas Kuhn,[4] has argued that these methods are influenced by prior beliefs and experiences. Consequently it cannot be expected that two scientists when observing, experiencing, or experimenting on the same event will make the same theory-neutral observations. The role of observation as a theory-neutral arbiter may not be possible. Theory-dependence of observation means that, even if there were agreed methods of inference and interpretation, scientists may still disagree on the nature of empirical data.[5]

15.2   See also

- Anecdotal evidence

- Empirical distribution function

- Empirical formula

- Empirical measure

- Empirical research (more on the scientific usage)

- Phenomenology (science)

- Scientific evidence

- Scientific method

- Theory

15.3   Footnotes

[1] Pickett 2006, p. 585

[2] Feldman 2001, p. 293

[3] Craig 2005, p. 1

[4] Kuhn 1970

[5] Bird 2013

15.4   References

- Bird, Alexander (2013). Zalta, Edward N., ed. "Thomas Kuhn". *Stanford Encyclopedia of Philosophy*. Section 4.2 Perception, Observational Incommensurability, and World-Change. Retrieved 25 January 2012.

- Craig, Edward (2005). "a posteriori". *The Shorter Routledge Encyclopedia of Philosophy*. Routledge. ISBN 9780415324953.

- Feldman, Richard (2001) [1999]. "Evidence". In Audi, Robert. *The Cambridge Dictionary of Philosophy* (2nd ed.). Cambridge, UK: Cambridge University Press. pp. 293–294. ISBN 978-0521637220.

- Kuhn, Thomas S. (1970) [1962]. *The Structure of Scientific Revolutions* (2nd ed.). Chicago: University of Chicago Press. ISBN 978-0226458045.

- Pickett, Joseph P., ed. (2011). "Empirical". *The American Heritage Dictionary of the English Language* (5th ed.). Houghton Mifflin. ISBN 978-0-547-04101-8.

15.5 External links

- The dictionary definition of empirical at Wiktionary
- The dictionary definition of evidence at Wiktionary
- A Priori and A Posteriori entry in the *Internet Encyclopedia of Philosophy*

Chapter 16

Mathematical induction

Mathematical induction can be informally illustrated by reference to the sequential effect of falling dominoes.

Mathematical induction is a method of mathematical proof typically used to establish a given statement for all natural numbers. It is a form of direct proof, and it is done in two steps. The first step, known as the **base case**, is to prove the given statement for the first natural number. The second step, known as the **inductive step**, is to prove that the given statement for any one natural number implies the given statement for the next natural number. From these two steps, mathematical induction is the rule from which we infer that the given statement is established for all natural numbers.

The method can be extended to prove statements about more general well-founded structures, such as trees; this generalization, known as structural induction, is used in mathematical logic and computer science. Mathematical induction in this extended sense is closely related to recursion. Mathematical induction, in some form, is the foundation of all correctness proofs for computer programs.[1]

Although its name may suggest otherwise, mathematical induction should not be misconstrued as a form of inductive reasoning (also see Problem of induction). Mathematical induction is an inference rule used in proofs. In mathematics, proofs including those using mathematical induction are examples of deductive reasoning, and inductive reasoning is excluded from proofs.[2]

16.1 History

In 370 BC, Plato's Parmenides may have contained an early example of an implicit inductive proof.[3] The earliest implicit traces of mathematical induction can be found in Euclid's[4][5][6] proof that the number of primes is infinite and in Bhaskara's "cyclic method".[7] An opposite iterated technique, counting *down* rather than up, is found in the Sorites paradox, where one argued that if 1,000,000 grains of sand formed a heap, and removing one grain from a heap left it a heap, then a single grain of sand (or even no grains) forms a heap.

An implicit proof by mathematical induction for arithmetic sequences was introduced in the *al-Fakhri* written by al-Karaji around 1000 AD, who used it to prove the binomial theorem and properties of Pascal's triangle.[8]

None of these ancient mathematicians, however, explicitly stated the inductive hypothesis. Another similar case (contrary to what Vacca has written, as Freudenthal carefully showed) was that of Francesco Maurolico in his *Arithmeticorum libri duo* (1575), who used the technique to prove that the sum of the first n odd integers is n^2. The first explicit formulation of the principle of induction was given by Pascal in his *Traité du triangle arithmétique* (1665). Another Frenchman, Fermat, made ample use of a related principle, indirect proof by infinite descent. The inductive hypothesis was also employed by the Swiss Jakob Bernoulli, and from then on it became more or less well known. The modern rigorous and systematic treatment of the principle came only in the 19th century, with George Boole,[9] Augustus de Morgan, Charles Sanders Peirce,[10][11] Giuseppe Peano, and Richard Dedekind.[7]

16.2 Description

The simplest and most common form of mathematical induction infers that a statement involving a natural number n holds for all values of n. The proof consists of two steps:

1. The **basis** (**base case**): prove that the statement holds for the first natural number n. Usually, $n = 0$ or $n = 1$, rarely, $n = -1$ (although not a natural number, the extension of the natural numbers to -1 is still a well-ordered set).

2. The **inductive step**: prove that, if the statement holds for some natural number n, then the statement holds for $n + 1$.

The hypothesis in the inductive step that the statement holds for some n is called the **induction hypothesis** (or **inductive hypothesis**). To perform the inductive step, one assumes the induction hypothesis and then uses this assumption to prove the statement for $n + 1$.

Whether $n = 0$ or $n = 1$ depends on the definition of the natural numbers. If 0 is considered a natural number, as is common in the fields of combinatorics and mathematical logic, the base case is given by $n = 0$. If, on the other hand, 1 is taken as the first natural number, then the base case is given by $n = 1$.

16.3 Example

Mathematical induction can be used to prove that the following statement, which we will call $P(n)$, holds for all natural numbers n.

$$0 + 1 + 2 + \cdots + n = \frac{n(n+1)}{2} \, .$$

$P(n)$ gives a formula for the sum of the natural numbers less than or equal to number n. The proof that $P(n)$ is true for each natural number n proceeds as follows.

Basis: Show that the statement holds for $n = 0$.
$P(0)$ amounts to the statement:

$$0 = \frac{0 \cdot (0+1)}{2} \, .$$

In the left-hand side of the equation, the only term is 0, and so the left-hand side is simply equal to 0. In the right-hand side of the equation, $0 \cdot (0 + 1)/2 = 0$.
The two sides are equal, so the statement is true for $n = 0$. Thus it has been shown that $P(0)$ holds.

Inductive step: Show that *if* $P(k)$ holds, then also $P(k + 1)$ holds. This can be done as follows.

Assume $P(k)$ holds (for some unspecified value of k). It must then be shown that $P(k + 1)$ holds, that is:

$$(0 + 1 + 2 + \cdots + k) + (k + 1) = \frac{(k+1)((k+1)+1)}{2} \, .$$

Using the induction hypothesis that $P(k)$ holds, the left-hand side can be rewritten to:

$$\frac{k(k+1)}{2} + (k+1) \, .$$

Algebraically:

$$\begin{aligned}
\frac{k(k+1)}{2} + (k+1) &= \frac{k(k+1) + 2(k+1)}{2} \\
&= \frac{(k+1)(k+2)}{2} \\
&= \frac{(k+1)((k+1)+1)}{2}
\end{aligned}$$

thereby showing that indeed $P(k + 1)$ holds.

Since both the basis and the inductive step have been performed, by mathematical induction, the statement $P(n)$ holds for all natural n. Q.E.D.

16.4   Axiom of induction

Mathematical induction as an inference rule can be formalized as a second-order axiom. The *axiom of induction* is, in logical symbols,

$$\forall P. \left[[P(0) \wedge \forall (k \in \mathbb{N}). [P(k) \Rightarrow P(k+1)]] \Rightarrow \forall (n \in \mathbb{N}). P(n)\right]$$

where P is any predicate and k and n are both natural numbers.

In words, the basis $P(0)$ and the inductive step (namely, that the inductive hypothesis $P(k)$ implies $P(k + 1)$) together imply that $P(n)$ for any natural number n. The axiom of induction asserts that the validity of inferring that $P(n)$ holds for any natural number n from the basis and the inductive step.

Note that the first quantifier in the axiom ranges over *predicates* rather than over individual numbers. This is a second-order quantifier, which means that this axiom is stated in second-order logic. Axiomatizing arithmetic induction in first-order logic requires an axiom schema containing a separate axiom for each possible predicate. The article Peano axioms contains further discussion of this issue.

16.4.1   Characterizing the structure of N by the induction axiom

Having proven the base case and the inductive step, then the structure of $\mathbb{N}$ is such that any value can be obtained by performing the inductive step repeatedly. It may be helpful to think of the domino effect. Consider a half line of dominoes each standing on end, and extending infinitely to the right (see picture). Suppose that:

1. The first domino falls right.

2. If a (fixed but arbitrary) domino falls right, then its next neighbor also falls right.

With these assumptions one can conclude (using mathematical induction) that all of the dominoes will fall right.

If the dominoes are arranged in another way, this conclusion needn't hold (see Peano axioms#Formulation for a counter example). Similarly, the induction axiom describes an essential property of $\mathbb{N}$, viz. that each of its members can be reached from 0 by sufficiently often adding 1. While there is only one structure that satisfies all Peano axioms (including induction),[12] there is no set of only first-order axioms that fulfils the same task.[13]

16.5   Variants

In practice, proofs by induction are often structured differently, depending on the exact nature of the property to be proved.

16.5.1   Induction basis other than 0 or 1

If we want to prove a statement not for all natural numbers but only for all numbers greater than or equal to a certain number b then the proof by induction consists of:

1. Showing that the statement holds when $n = b$.

2. Showing that if the statement holds for $n = m \geq b$ then the same statement also holds for $n = m + 1$.

This can be used, for example, to show that $n^2 \geq 3n$ for $n \geq 3$. A more substantial example is a proof that

$$\frac{n^n}{3^n} < n! < \frac{n^n}{2^n} \text{ for } n \geq 6.$$

In this way we can prove that $P(n)$ holds for all $n \geq 1$, or even $n \geq -5$. This form of mathematical induction is actually a special case of the previous form because if the statement that we intend to prove is $P(n)$ then proving it with these two rules is equivalent with proving $P(n + b)$ for all natural numbers n with the first two steps.

16.5.2 Induction basis equal to 2

In mathematics, many standard functions, including operations such as "+" and relations such as "=", are binary, meaning that they take two arguments. Often these functions possess properties that implicitly extend them to more than two arguments. For example, once addition $a + b$ is defined and is known to satisfy the associativity property $(a + b) + c = a + (b + c)$, then the ternary addition $a + b + c$ makes sense, either as $(a + b) + c$ or as $a + (b + c)$. Similarly, many axioms and theorems in mathematics are stated only for the binary versions of mathematical operations and relations, and implicitly extend to higher-arity versions.

Suppose that we wish to prove a statement about an n-ary operation implicitly defined from a binary operation, using mathematical induction on n. In this case it is natural to take 2 for the induction basis.

Example: product rule for the derivative

In this example, the binary operation in question is multiplication (of functions). The usual product rule for the derivative taught in calculus states:

$$(fg)' = f'g + g'f.$$

or in logarithmic derivative form

$$(fg)'/(fg) = f'/f + g'/g.$$

This can be generalized to a product of n functions. One has

$$(f_1 f_2 f_3 \cdots f_n)'$$

$$= (f_1' f_2 f_3 \cdots f_n) + (f_1 f_2' f_3 \cdots f_n) + (f_1 f_2 f_3' \cdots f_n) + \cdots + (f_1 f_2 \cdots f_{n-1} f_n').$$

or in logarithmic derivative form

$$(f_1 f_2 f_3 \cdots f_n)'/(f_1 f_2 f_3 \cdots f_n)$$

$$= (f_1'/f_1) + (f_2'/f_2) + (f_3'/f_3) + \cdots + (f_n'/f_n).$$

In each of the n terms of the usual form, just one of the factors is a derivative; the others are not.

When this general fact is proved by mathematical induction, the $n = 0$ case is trivial, $(1)' = 0$ (since the empty product is 1, and the empty sum is 0). The $n = 1$ case is also trivial, $f_1' = f_1'$. And for each $n \geq 3$, the case is easy to prove from the preceding $n - 1$ case. The real difficulty lies in the $n = 2$ case, which is why that is the one stated in the standard product rule.

16.5.3 Induction on more than one counter

It is sometimes desirable to prove a statement involving two natural numbers, n and m, by iterating the induction process. That is, one performs a basis step and an inductive step for n, and in each of those performs a basis step and an inductive step for m. See, for example, the proof of commutativity accompanying *addition of natural numbers*. More complicated arguments involving three or more counters are also possible.

16.5.4 Infinite descent

Main article: Infinite descent

The method of infinite descent was one of Pierre de Fermat's favorites. This method of proof can assume several slightly different forms. For example, it might begin by showing that if a statement is true for a natural number n it must also be true for some smaller natural number m ($m < n$). Using mathematical induction (implicitly) with the inductive hypothesis being that the statement is false for all natural numbers less than or equal to m, we can conclude that the statement cannot be true for any natural number n.

Although this particular form of infinite-descent proof is clearly a mathematical induction, whether one holds all proofs "by infinite descent" to be mathematical inductions depends on how one defines the term "proof by infinite descent." One might, for example, use the term to apply to proofs in which the well-ordering of the natural numbers is assumed, but not the principle of induction. Such, for example, is the usual proof that 2 has no rational square root (see Infinite descent).

16.5.5 Prefix induction

The most common form of induction requires proving that

$$\forall k(P(k) \rightarrow P(k+1))$$

or equivalently

$$\forall k(P(k-1) \rightarrow P(k))$$

whereupon the induction principle "automates" n applications of this inference in getting from $P(0)$ to $P(n)$. This could be called "predecessor induction" because each step proves something about a number from something about that number's predecessor.

A variant of interest in computational complexity is "prefix induction", in which one needs to prove

$$\forall k(P(k) \rightarrow P(2k) \wedge P(2k+1))$$

or equivalently

$$\forall k \left(P\left(\left\lfloor \frac{k}{2} \right\rfloor \right) \rightarrow P(k) \right)$$

The induction principle then "automates" log n applications of this inference in getting from $P(0)$ to $P(n)$. (It is called "prefix induction" because each step proves something about a number from something about the "prefix" of that number formed by truncating the low bit of its binary representation.)

If traditional predecessor induction is interpreted computationally as an n-step loop, prefix induction corresponds to a log n-step loop, and thus proofs using prefix induction are "more feasibly constructive" than proofs using predecessor induction.

Predecessor induction can trivially simulate prefix induction on the same statement. Prefix induction can simulate predecessor induction, but only at the cost of making the statement more syntactically complex (adding a bounded universal quantifier), so the interesting results relating prefix induction to polynomial-time computation depend on excluding unbounded quantifiers entirely, and limiting the alternation of bounded universal and existential quantifiers allowed in the statement. See [14]

One could take it a step farther to "prefix of prefix induction": one must prove

$$\forall k \left(P\left(\left\lfloor\sqrt{k}\right\rfloor\right) \to P(k) \right)$$

whereupon the induction principle "automates" log log n applications of this inference in getting from $P(0)$ to $P(n)$. This form of induction has been used, analogously, to study log-time parallel computation.

16.5.6   Complete induction

Another variant, called **complete induction** (or **strong induction** or **course of values induction**), says that in the second step we may assume not only that the statement holds for $n = m$ but also that it is true for **all** n less than or equal to m.

Complete induction is most useful when several instances of the inductive hypothesis are required for each inductive step. For example, complete induction can be used to show that

$$F_n = \frac{\varphi^n - \psi^n}{\varphi - \psi}$$

where Fn is the nth Fibonacci number, $\varphi = (1 + \sqrt{5})/2$ (the golden ratio) and $\psi = (1 - \sqrt{5})/2$ are the roots of the polynomial $x^2 - x - 1$. By using the fact that $Fn + 2 = Fn + 1 + Fn$ for each $n \in \mathbf{N}$, the identity above can be verified by direct calculation for $Fn + 2$ if we assume that it already holds for both $Fn + 1$ and Fn. To complete the proof, the identity must be verified in the two base cases $n = 0$ and $n = 1$.

Another proof by complete induction uses the hypothesis that the statement holds for *all* smaller n more thoroughly. Consider the statement that "every natural number greater than 1 is a product of (one or more) prime numbers", and assume that for a given $m > 1$ it holds for all smaller $n > 1$. If m is prime then it is certainly a product of primes, and if not, then by definition it is a product: $m = n_1 \, n_2$, where neither of the factors is equal to 1; hence neither is equal to m, and so both are smaller than m. The induction hypothesis now applies to n_1 and n_2, so each one is a product of primes. Then m is a product of products of primes; i.e. a product of primes.

This generalization, complete induction, is equivalent to the ordinary mathematical induction described above. Suppose $P(n)$ is the statement that we intend to prove by complete induction. Let $Q(n)$ mean $P(m)$ holds for all m such that $0 \le m \le n$. Then $Q(n)$ is true for all n if and only if $P(n)$ is true for all n, and a proof of $P(n)$ by complete induction is just the same thing as a proof of $Q(n)$ by (ordinary) induction.

Transfinite induction

Main article: Transfinite induction

The last two steps can be reformulated as one step:

  1. Showing that if the statement holds for all $n < m$ then the same statement also holds for $n = m$.

This form of mathematical induction is not only valid for statements about natural numbers, but for statements about elements of any well-founded set, that is, a set with an irreflexive relation $<$ that contains no infinite descending chains.

This form of induction, when applied to ordinals (which form a well-ordered and hence well-founded class), is called *transfinite induction*. It is an important proof technique in set theory, topology and other fields.

Proofs by transfinite induction typically distinguish three cases:

  1. when m is a minimal element, i.e. there is no element smaller than m

2. when m has a direct predecessor, i.e. the set of elements which are smaller than m has a largest element

3. when m has no direct predecessor, i.e. m is a so-called limit-ordinal

Strictly speaking, it is not necessary in transfinite induction to prove the basis, because it is a vacuous special case of the proposition that if P is true of all $n < m$, then P is true of m. It is vacuously true precisely because there are no values of $n < m$ that could serve as counterexamples.

16.6 Equivalence with the well-ordering principle

The principle of mathematical induction is usually stated as an axiom of the natural numbers; see Peano axioms. However, it can be proved from the well-ordering principle. Indeed, suppose the following:

- The set of natural numbers is well-ordered.

- Every natural number is either zero, or $n+1$ for some natural number n.

- For any natural number n, $n+1$ is greater than n.

To derive simple induction from these axioms, we must show that if P(n) is some proposition predicated of n, and if:

- P(0) holds and

- whenever P(k) is true then P($k+1$) is also true

then P(n) holds for all n.

Proof. Let S be the set of all natural numbers for which P(n) is false. Let us see what happens if we assert that S is nonempty. Well-ordering tells us that S has a least element, say t. Moreover, since P(0) is true, t is not 0. Since every natural number is either zero or some $n+1$, there is some natural number n such that $n+1=t$. Now n is less than t, and t is the least element of S. It follows that n is not in S, and so P(n) is true. This means that P($n+1$) is true, and so P(t) is true. This is a contradiction, since t was in S. Therefore, S is empty.

It can also be proved that induction, given the other axioms, implies the well-ordering principle.

16.7 Example of error in the inductive step

Main article: All horses are the same color

This example demonstrated a subtle error in the proof of the inductive step.

Joel E. Cohen proposed the following argument, which purports to prove by mathematical induction that all horses are of the same color:[15]

- Basis: If there is only *one* horse, there is only one color.

- Induction step: Assume as induction hypothesis that within any set of n horses, there is only one color. Now look at any set of $n + 1$ horses. Number them: 1, 2, 3, ..., n, $n + 1$. Consider the sets {1, 2, 3, ..., n} and {2, 3, 4, ..., $n + 1$}. Each is a set of only n horses, therefore within each there is only one color. But the two sets overlap, so there must be only one color among all $n + 1$ horses.

The basis case $n = 1$ is trivial (as any horse is the same color as itself), and the inductive step is correct in all cases $n > 1$. However, the logic of the inductive step is incorrect for $n = 1$, because the statement that "the two sets overlap" is false (there are only $n + 1 = 2$ horses prior to either removal, and after removal the sets of one horse each do not overlap).

16.8   See also

- Combinatorial proof

- Recursion

- Recursion (computer science)

- Structural induction

16.9   Notes

[1]   Anderson, Robert B. (1979). *Proving Programs Correct*. New York: John Wiley & Sons. p. 1. ISBN 0471033952.

[2]   Suber, Peter. "Mathematical Induction". Earlham College. Retrieved 26 March 2011.

[3]   Mathematical Induction: The Basis Step of Verification and Validation in a Modeling and Simulation Course

[4]   Chris K. Caldwell. "Euclid's Proof of the Infinitude of Primes (c. 300 BC)". *utm.edu*.

[5]   "Euclid's Primes". *mathsisgoodforyou.com*.

[6]   "Proofs of the Infinity of the Prime Numbers". *hermetic.ch*.

[7]   Cajori (1918), p. 197: 'The process of reasoning called "Mathematical Induction" has had several independent origins. It has been traced back to the Swiss Jakob (James) Bernoulli, the Frenchman B. Pascal and P. Fermat, and the Italian F. Maurolycus. [...] By reading a little between the lines one can find traces of mathematical induction still earlier, in the writings of the Hindus and the Greeks, as, for instance, in the "cyclic method" of Bhaskara, and in Euclid's proof that the number of primes is infinite.'

[8]   Rashed, R. (1994), "Mathematical induction: al-Karajī and al-Samaw'al", *The Development of Arabic Mathematics: Between Arithmetic and Algebra*, Boston Studies in the Philosophy of Science **156**, Kluwer Academic Publishers, pp. 62–84, ISBN 9780792325659

[9]   "It is sometimes required to prove a theorem which shall be true whenever a certain quantity *n* which it involves shall be an integer or whole number and the method of proof is usually of the following kind. *1st*. The theorem is proved to be true when *n* = 1. *2ndly*. It is proved that if the theorem is true when *n* is a given whole number, it will be true if *n* is the next greater integer. Hence the theorem is true universally. . . .. This species of argument may be termed a continued *sorites*" (Boole circa 1849 *Elementary Treatise on Logic not mathematical* pages 40–41 reprinted in Grattan-Guinness, Ivor and Bornet, Gérard (1997), *George Boole: Selected Manuscripts on Logic and its Philosophy*, Birkhäuser Verlag, Berlin, ISBN 3-7643-5456-9)

[10] Peirce, C. S. (1881). "On the Logic of Number". *American Journal of Mathematics* **4** (1–4). pp. 85–95. doi:10.2307/2369151. JSTOR 2369151. MR 1507856. Reprinted (CP 3.252-88), (W 4:299-309).

[11] Shields (1997)

[12] Hermes (1973), VI.3.1

[13] Hermes (1973), VI.4.3, presenting a theorem of Thoralf Skolem

[14] Buss, Samuel (1986). *Bounded Arithmetic*. Naples: Bibliopolis.

[15] Cohen, Joel E. (1961), "On the nature of mathematical proof", *Opus*. Reprinted in *A Random Walk in Science* (R. L. Weber, ed.), Crane, Russak & Co., 1973.

16.10 References

16.10.1 Introduction

- Franklin, J.; A. Daoud (2011). *Proof in Mathematics: An Introduction*. Sydney: Kew Books. ISBN 0-646-54509-4. (Ch. 8.)

- Hazewinkel, Michiel, ed. (2001), "Mathematical induction", *Encyclopedia of Mathematics*, Springer, ISBN 978-1-55608-010-4

- Hermes, Hans (1973). *Introduction to Mathematical Logic*. Hochschultext. London: Springer. ISBN 3540058192. ISSN 1431-4657.

- Knuth, Donald E. (1997). *The Art of Computer Programming, Volume 1: Fundamental Algorithms* (3rd ed.). Addison-Wesley. ISBN 0-201-89683-4. (Section 1.2.1: Mathematical Induction, pp. 11–21.)

- Kolmogorov, Andrey N.; Sergei V. Fomin (1975). *Introductory Real Analysis*. Silverman, R. A. (trans., ed.). New York: Dover. ISBN 0-486-61226-0. (Section 3.8: Transfinite induction, pp. 28–29.)

16.10.2 History

- Acerbi, F. (2000). "Plato: *Parmenides* 149a7-c3. A Proof by Complete Induction?". *Archive for History of Exact Sciences* **55**: 57–76. doi:10.1007/s004070000020.

- Bussey, W. H. (1917). "The Origin of Mathematical Induction". *The American Mathematical Monthly* **24** (5): 199–207. doi:10.2307/2974308. JSTOR 2974308.

- Cajori, Florian (1918). "Origin of the Name "Mathematical Induction"". *The American Mathematical Monthly* **25** (5): 197–201. doi:10.2307/2972638. JSTOR 2972638.

- Fowler D. (1994). "Could the Greeks Have Used Mathematical Induction? Did They Use It?". *Physis* **XXXI**: 253–265.

- Freudenthal, Hans (1953). "Zur Geschichte der vollständigen Induction". *Archives Internationales d'Histiore des Sciences* **6**: 17–37.

- Katz, Victor J. (1998). *History of Mathematics: An Introduction*. Addison-Wesley. ISBN 0-321-01618-1.

- Peirce, C. S. (1881). "On the Logic of Number". *American Journal of Mathematics* **4** (1–4). pp. 85–95. doi:10.2307/2369151. JSTOR 2369151. MR 1507856. Reprinted (CP 3.252-88), (W 4:299-309).

- Rabinovitch, Nachum L. (1970). "Rabbi Levi Ben Gershon and the origins of mathematical induction". *Archive for History of Exact Sciences* **6** (3): 237–248. doi:10.1007/BF00327237.

- Rashed, Roshdi (1972). "L'induction mathématique: al-Karajī, as-Samaw'al". *Archive for History of Exact Sciences* (in French) **9** (1): 1–21. doi:10.1007/BF00348537.

- Shields, Paul (1997). "Peirce's Axiomatization of Arithmetic". In Houser et al. *Studies in the Logic of Charles S. Peirce*.

- Ungure, S. (1991). "Greek Mathematics and Mathematical Induction". *Physis*. XXVIII: 273–289.

- Ungure, S. (1994). "Fowling after Induction". *Physis* **XXXI**: 267–272.

- Vacca, G. (1909). "Maurolycus, the First Discoverer of the Principle of Mathematical Induction". *Bulletin of the American Mathematical Society* **16** (2): 70–73. doi:10.1090/S0002-9904-1909-01860-9.

- Yadegari, Mohammad (1978). "The Use of Mathematical Induction by Abū Kāmil Shujā' Ibn Aslam (850-930)". *Isis* **69** (2): 259–262. doi:10.1086/352009. JSTOR 230435.

Chapter 17

Direct proof

In mathematics and logic, a **direct proof** is a way of showing the truth or falsehood of a given statement by a straight-forward combination of established facts, usually axioms, existing lemmas and theorems, without making any further assumptions.[1] In order to directly prove a conditional statement of the form "If p, then q", it suffices to consider the situations in which the statement p is true. Logical deduction is employed to reason from assumptions to conclusion. The type of logic employed is almost invariably first-order logic, employing the quantifiers *for all* and *there exists*. Common proof rules used are modus ponens and universal instantiation.[2]

In contrast, an indirect proof may begin with certain hypothetical scenarios and then proceed to eliminate the uncertainties in each of these scenarios until an inescapable conclusion is forced. For example instead of showing directly $p \Rightarrow q$, one proves its contrapositive $\sim q \Rightarrow \sim p$ (one assumes $\sim q$ and shows that it leads to $\sim p$). Since $p \Rightarrow q$ and $\sim q \Rightarrow \sim p$ are equivalent by the principle of transposition (see law of excluded middle), $p \Rightarrow q$ is indirectly proved. Proof methods that are not direct include proof by contradiction, including proof by infinite descent. Direct proof methods include proof by exhaustion and proof by induction.

17.1 History and etymology

A direct proof is the simplest form of proof there is. The word 'proof' comes from the Latin word probare,[3] which means "to test". The earliest use of proofs was prominent in legal proceedings. A person with authority, such as a nobleman, was said to have probity, which means that the evidence was by his relative authority, which outweighed empirical testimony. In days gone by, mathematics and proof was often intertwined with practical questions – with populations like the Egyptians and the Greeks showing an interest in surveying land.[4] This lead to a natural curiosity with regards to geometry and trigonometry – particularly triangles and rectangles. These were the shapes which provided the most questions in terms of practical things, so early geometrical concepts were focused on these shapes, for example, the likes of buildings and pyramids used these shapes in abundance. Another shape which is crucial in the history of direct proof is the circle, which was crucial for the design of arenas and water tanks. This meant that ancient geometry (and Euclidean Geometry) discussed circles.

The earliest form of mathematics was phenomenological. For example, if someone could draw a reasonable picture, or give a convincing description, then that met all the criteria for something to be described as a mathematical "fact". On occasion, analogical arguments took place, or even by "invoking the gods". The idea that mathematical statements could be proven had not been developed yet, so these were the earliest forms of the concept of proof, despite not being actual proof at all.

Proof as we know it came about with one specific question: "what is a proof?" Traditionally, a proof is a platform which convinces someone beyond reasonable doubt that a statement is mathematically true. Naturally, one would assume that the best way to prove the truth of something like this (B) would be to draw up a comparison with something old (A) that has already been proven as true. Thus was created the concept of deriving a new result from an old result.

Geometric Constructions

17.2 Examples

17.2.1 The sum of two even integers equals an even integer

Consider two even integers x and y. Since they are even, they can be written as

$$x = 2a$$

$$y = 2b$$

respectively for integers a and b. Then the sum can be written as

$$x + y = 2a + 2b = 2(a + b)$$

From this it is clear $x + y$ has 2 as a factor and therefore is even, so the sum of any two even integers is even.

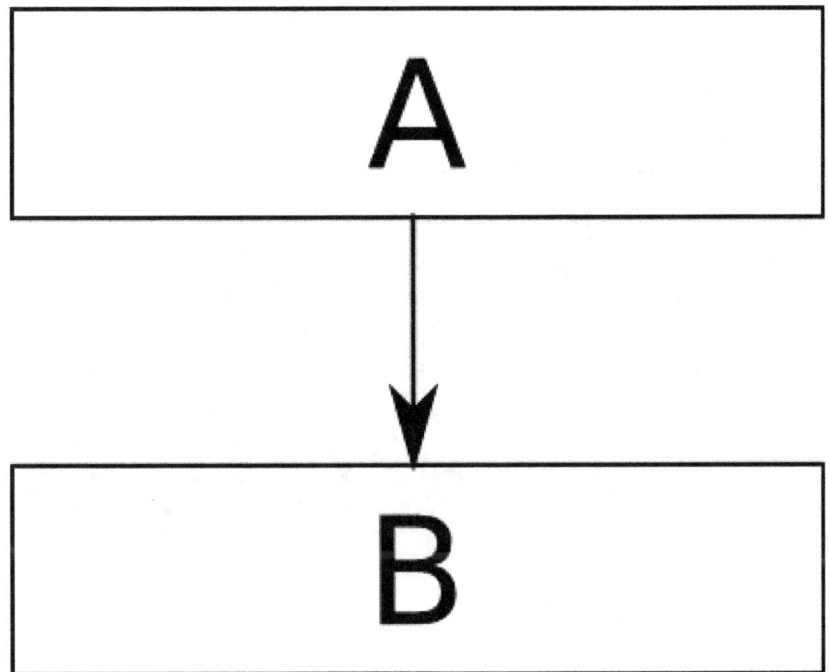

New result from an old result

17.2.2 Pythagoras' Theorem

Observe that we have four right-angled triangles and a square packed into a large square. Each of the triangles has sides a and b and hypotenuse c. The area of a square is defined as the square of the length of its sides - in this case, $(a + b)^2$. However, the area of the large square can also be expressed as the sum of the areas of its components. In this case, that would be the sum of the areas of the four triangles and the small square in the middle.[5]

We know that the area of the large square is equal to $(a + b)^2$

The area of a triangle is equal to $\frac{1}{2}ab$

We know that the area of the large square is also equal to the sum of the areas of the triangles, plus the area of the small square, and thus the area of the large square equals $4(\frac{1}{2}ab) + c^2$

These are equal, and so:

$$(a + b)^2 = 4(1/2ab) + c^2$$

After some simplifying:

$$a^2 + 2ab + b^2 = 2ab + c^2$$

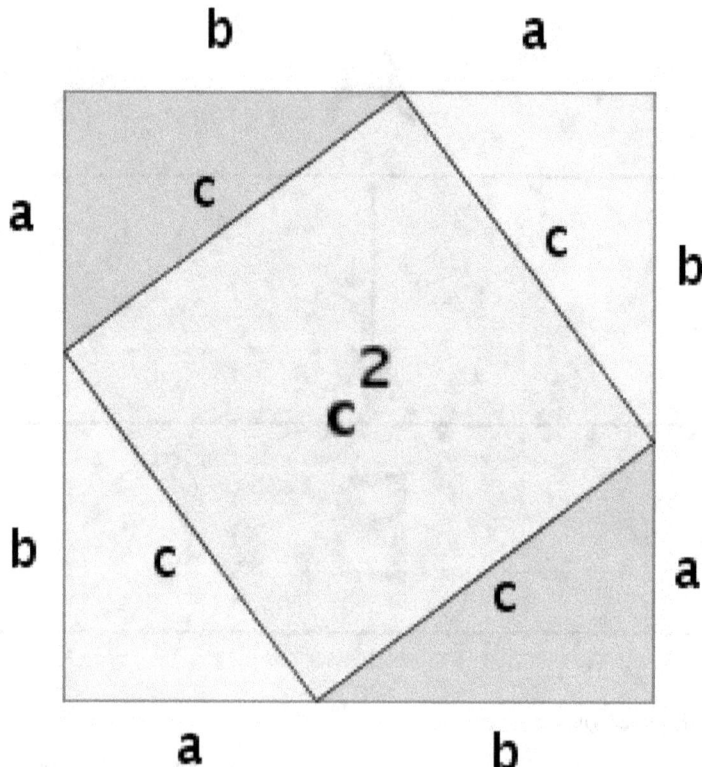

Diagram of Pythagoras Theorem

Removing the ab that appears on both sides gives

$$a^2 + b^2 = c^2$$

Which proves Pythagoras' theorem. ∎

17.2.3 If n is an odd integer, n^2 is also an odd integer.

By definition, if n is an odd integer, it can be expressed as:

$n = 2k + 1$

for some integer k. Thus:

$$
\begin{aligned}
n^2 &= (2k+1)^2 \\
&= (2k+1)(2k+1) \\
&= 4k^2 + 2k + 2k + 1 \\
&= 4k^2 + 4k + 1 \\
&= 2(2k^2 + 2k) + 1
\end{aligned}
$$

As $(2k^2 + 2k)$ is an integer, our answer can be expressed as:

$2k + 1$

And hence we have shown that n^2 is odd. ∎

17.3   References

[1] Cupillari, Antonella. *The Nuts and Bolts of Proofs*. Academic Press, 2001. Page 3.

[2] C. Gupta, S. Singh, S. Kumar *Advanced Discrete Structure*. I.K. International Publishing House Pvt. Ltd., 2010. Page 127.

[3] New Shorter Oxford English Dictionary

[4] Krantz, Steven G. *The History and Concept of Mathematical Proof*. February 5, 2007.

[5] Krantz, Steven G. *The Proof is the Pudding*. Springer, 2010. Page 43.

17.4   Sources

• Franklin, J.; A. Daoud (2011). *Proof in Mathematics: An Introduction*. Sydney: Kew Books. ISBN 0-646-54509-4. (Ch. 1.)

17.5   External links

• Direct Proof from Larry W. Cusick's How To Write Proofs.

• Direct Proofs from Patrick Keef and David Guichard's Introduction to Higher Mathematics.

• Direct Proof section of Richard Hammack's Book of Proof.

Chapter 18

Proof by contrapositive

In logic, the **contrapositive** of a conditional statement is formed by negating both terms and reversing the direction of inference. Explicitly, the contrapositive of the statement "if A, then B" is "if not B, then not A." A statement and its contrapositive are logically equivalent: if the statement is true, then its contrapositive is true, and vice versa.[1]

In mathematics, **proof by contraposition** is a rule of inference used in proofs. This rule infers a conditional statement from its contrapositive.[2] In other words, the conclusion "if A, then B" is drawn from the single premise "if not B, then not A."

18.1   Example

Let x be an integer.

To prove: *If* x^2 *is even, then* x *is even.*

Although a direct proof can be given, we choose to prove this statement by contraposition. The contrapositive of the above statement is:

If x *is not even, then* x^2 *is not even.*

This latter statement can be proven as follows. Suppose x is not even. Then x is odd. The product of two odd numbers is odd, hence $x^2 = x \cdot x$ is odd. Thus x^2 is not even.

Having proved the contrapositive, we infer the original statement.[3]

18.2   Relation to proof by contradiction

Any proof by contrapositive can also be trivially formulated in terms of a Proof by contradiction: To prove the proposition $P \Rightarrow Q$, we consider the opposite, $\neg(P \Rightarrow Q) \equiv \neg(\neg P \vee Q) \equiv P \wedge \neg Q$. Since we have a proof that $\neg Q \Rightarrow \neg P$, we have $P \wedge \neg Q \Rightarrow P \wedge \neg P \equiv \bot$ which arrives at the contradiction we want. So proof by contrapositive is in some sense "at least as hard to formulate" as proof by contradiction.

18.3   See also

• Contraposition

- Modus tollens
- *Reductio ad absurdum*

18.4 References

[1] Regents Exam Prep, contrapositive definition

[2] Larry Cusick's (CSU-Fresno) How to write proofs tutorial

[3] Franklin, J.; A. Daoud (2011). *Proof in Mathematics: An Introduction*. Sydney: Kew Books. ISBN 0-646-54509-4. (p. 50).

Chapter 19

Proof by contradiction

See also: Table of logic symbols

In logic, **proof by contradiction** is a form of proof, and more specifically a form of indirect proof, that establishes the truth or validity of a proposition by showing that the proposition's being false would imply a contradiction. Proof by contradiction is also known as **indirect proof**, **apagogical argument**, **proof by assuming the opposite**, and *reductio ad impossibilem*. It is a particular kind of the more general form of argument known as *reductio ad absurdum*.

G. H. Hardy described proof by contradiction as "one of a mathematician's finest weapons", saying "It is a far finer gambit than any chess gambit: a chess player may offer the sacrifice of a pawn or even a piece, but a mathematician offers the game."[1]

19.1   Examples

19.1.1   Irrationality of the square root of 2

A classic proof by contradiction from mathematics is the proof that the square root of 2 is irrational.[2] If it were rational, it could be expressed as a fraction a/b in lowest terms, where a and b are integers, at least one of which is odd. But if $a/b = \sqrt{2}$, then $a^2 = 2b^2$. Therefore a^2 must be even. Because the square of an odd number is odd, that in turn implies that a is even. This means that b must be odd because a/b is in lowest terms.

On the other hand, if a is even, then a^2 is a multiple of 4. If a^2 is a multiple of 4 and $a^2 = 2b^2$, then $2b^2$ is a multiple of 4, and therefore b^2 is even, and so is b.

So b is odd and even, a contradiction. Therefore the initial assumption—that $\sqrt{2}$ can be expressed as a fraction—must be false.

19.1.2   The length of the hypotenuse

The method of proof by contradiction has also been used to show that for any non-degenerate right triangle, the length of the hypotenuse is less than the sum of the lengths of the two remaining sides.[3] The proof relies on the Pythagorean theorem. Letting c be the length of the hypotenuse and a and b the lengths of the legs, the claim is that $a + b > c$.

The claim is negated to assume that $a + b \leq c$. Squaring both sides results in $(a + b)^2 \leq c^2$ or, equivalently, $a^2 + 2ab + b^2 \leq c^2$. A triangle is non-degenerate if each edge has positive length, so it may be assumed that a and b are greater than 0. Therefore, $a^2 + b^2 < a^2 + 2ab + b^2 \leq c^2$. The transitive relation may be reduced to $a^2 + b^2 < c^2$. It is known from the Pythagorean theorem that $a^2 + b^2 = c^2$. This results in a contradiction since strict inequality and equality are mutually exclusive. The latter was a result of the Pythagorean theorem and the former the assumption that $a + b \leq c$.

The contradiction means that it is impossible for both to be true and it is known that the Pythagorean theorem holds. It follows that the assumption that $a + b \leq c$ must be false and hence $a + b > c$, proving the claim.

19.1.3 No least positive rational number

Consider the proposition, P: "there is no smallest rational number greater than 0". In a proof by contradiction, we start by assuming the opposite, $\neg P$: that there *is* a smallest rational number, say, r.

Now $r/2$ is a rational number greater than 0 and smaller than r. (In the above symbolic argument, "$r/2$ is the smallest rational number" would be Q and "r (which is different from $r/2$) is the smallest rational number" would be $\neg Q$.) But that contradicts our initial assumption, $\neg P$, that r was the *smallest* rational number. So we can conclude that the original proposition, P, must be true — "there is no smallest rational number greater than 0".

19.1.4 Other

For other examples, see proof that the square root of 2 is not rational (where indirect proofs different from the above one can be found) and Cantor's diagonal argument.

19.2   In mathematical logic

In mathematical logic, the proof by contradiction is represented as:

If

$S \cup \{P\} \vdash \mathbb{F}$

then

$S \vdash \neg P.$

or

If

$S \cup \{\neg P\} \vdash \mathbb{F}$

then

$S \vdash P.$

In the above, P is the proposition we wish to prove, and S is a set of statements, which are the premises—these could be, for example, the axioms of the theory we are working in, or earlier theorems we can build upon. We consider P, or the negation of P, in addition to S; if this leads to a logical contradiction F, then we can conclude that the statements in S lead to the negation of P, or P itself, respectively.

Note that the set-theoretic union, in some contexts closely related to logical disjunction (or), is used here for sets of statements in such a way that it is more related to logical conjunction (and).

A particular kind of indirect proof assumes that some object doesn't exist, and then proves that this would lead to a contradiction; thus, such an object must exist. Although it is quite freely used in mathematical proofs, not every school of mathematical thought accepts this kind of argument as universally valid. See further Nonconstructive proof.

19.3 Notation

Proofs by contradiction sometimes end with the word "Contradiction!". Isaac Barrow and Baermann used the notation Q.E.A., for "*quod est absurdum*" ("which is absurd"), along the lines of Q.E.D., but this notation is rarely used today.[4] A graphical symbol sometimes used for contradictions is a downwards zigzag arrow "lightning" symbol (U+21AF: ↯), for example in Davey and Priestley.[5] Others sometimes used include a pair of opposing arrows (as →←— or ⇒⇐), struck-out arrows (↮), a stylized form of hash (such as U+2A33: ⨳), or the "reference mark" (U+203B: ※).[6][7] The "up tack" symbol (U+22A5: ⊥) used by philosophers and logicians (see contradiction) also appears, but is often avoided due to its usage for orthogonality.

19.4 See also

- Proof by contrapositive

19.5 References

[1] G. H. Hardy, *A Mathematician's Apology; Cambridge University Press, 1992. ISBN 9780521427067.* p. 94.

[2] Alfeld, Peter (16 August 1996). "Why is the square root of 2 irrational?". *Understanding Mathematics, a study guide.* Department of Mathematics, University of Utah. Retrieved 6 February 2013.

[3] Stone, Peter. "Logic, Sets, and Functions: Honors" (PDF). *Course materials.* pp 14–23: Department of Computer Sciences, The University of Texas at Austin. Retrieved 6 February 2013.

[4] Hartshorne on QED and related

[5] B. Davey and H.A. Priestley, Introduction to lattices and order, Cambridge University Press, 2002.

[6] The Comprehensive LaTeX Symbol List, pg. 20. http://www.ctan.org/tex-archive/info/symbols/comprehensive/symbols-a4.pdf

[7] Gary Hardegree, *Introduction to Modal Logic*, Chapter 2, pg. II–2. http://people.umass.edu/gmhwww/511/pdf/c02.pdf

19.6 Further reading

- Franklin, James (2011). *Proof in Mathematics: An Introduction.* chapter 6: Kew. ISBN 978-0-646-54509-7.

19.7 External links

- Proof by Contradiction from Larry W. Cusick's How To Write Proofs

Chapter 20

Constructive proof

In mathematics, a **constructive proof** is a method of proof that demonstrates the existence of a mathematical object by creating or providing a method for creating the object. This is in contrast to a **non-constructive proof** (also known as an *existence proof* or *pure existence theorem*) which proves the existence of a particular kind of object without providing an example.

Some non-constructive proofs show that if a certain proposition is false, a contradiction ensues; consequently the proposition must be true (proof by contradiction). However, the principle of explosion (*ex falso quodlibet*) has been accepted in some varieties of constructive mathematics, including intuitionism.

Constructivism is a mathematical philosophy that rejects all but constructive proofs in mathematics. This leads to a restriction on the proof methods allowed (prototypically, the law of the excluded middle is not accepted) and a different meaning of terminology (for example, the term "or" has a stronger meaning in constructive mathematics than in classical).

Constructive proofs can be seen as defining certified mathematical algorithms: this idea is explored in the Brouwer–Heyting–Kolmogorov interpretation of constructive logic, the Curry–Howard correspondence between proofs and programs, and such logical systems as Per Martin-Löf's Intuitionistic Type Theory, and Thierry Coquand and Gérard Huet's Calculus of Constructions.

20.1 Examples

20.1.1 Non-constructive proofs

First consider the theorem that there are an infinitude of prime numbers. Euclid's proof is constructive. But a common way of simplifying Euclid's proof postulates that, contrary to the assertion in the theorem, there are only a finite number of them, in which case there is a largest one, denoted n. Then consider the number $n! + 1$ (1 + the product of the first n numbers). Either this number is prime, or all of its prime factors are greater than n. Without establishing a specific prime number, this proves that one exists that is greater than n, contrary to the original postulate.

Now consider the theorem "There exist irrational numbers a and b such that a^b is rational." This theorem can be proven using a constructive proof, or using a non-constructive proof.

The following 1953 proof by Dov Jarden has been widely used as an example of a non-constructive proof since at least 1970:[1][2]

> **CURIOSA**
> **339.** *A Simple Proof That a Power of an Irrational Number to an Irrational Exponent May Be Rational.*
> $\sqrt{2}^{\sqrt{2}}$ is either rational or irrational. If it is rational, our statement is proved. If it is irrational, $(\sqrt{2}^{\sqrt{2}})^{\sqrt{2}} = 2$
> proves our statement.
> Dov Jarden Jerusalem

In a bit more detail:

- Recall that $\sqrt{2}$ is irrational, and 2 is rational. Consider the number $q = \sqrt{2}^{\sqrt{2}}$. Either it is rational or it is irrational.
- If q is rational, then the theorem is true, with a and b both being $\sqrt{2}$.
- If q is irrational, then the theorem is true, with a being $\sqrt{2}^{\sqrt{2}}$ and b being $\sqrt{2}$, since

$$\left(\sqrt{2}^{\sqrt{2}}\right)^{\sqrt{2}} = \sqrt{2}^{(\sqrt{2} \cdot \sqrt{2})} = \sqrt{2}^2 = 2.$$

This proof is non-constructive because it relies on the statement "Either q is rational or it is irrational"—an instance of the law of excluded middle, which is not valid within a constructive proof. The non-constructive proof does not construct an example a and b; it merely gives a number of possibilities (in this case, two mutually exclusive possibilities) and shows that one of them—but does not show *which* one—must yield the desired example.

(It turns out that $\sqrt{2}^{\sqrt{2}}$ is irrational because of the Gelfond–Schneider theorem, but this fact is irrelevant to the correctness of the non-constructive proof.)

20.1.2 Constructive proofs

A *constructive* proof of the above theorem on irrational powers of irrationals would give an actual example, such as:

$$a = \sqrt{2}, \quad b = \log_2 9, \quad a^b = 3.$$

The square root of 2 is irrational, and 3 is rational. $\log_2 9$ is also irrational: if it were equal to $\frac{m}{n}$, then, by the properties of logarithms, 9^n would be equal to 2^m, but the former is odd, and the latter is even.

A more substantial example is the graph minor theorem. A consequence of this theorem is that a graph can be drawn on the torus if, and only if, none of its minors belong to a certain finite set of "forbidden minors". However, the proof of the existence of this finite set is not constructive, and the forbidden minors are not actually specified. They are still unknown.

20.2 Brouwerian counterexamples

In constructive mathematics, a statement may be disproved by giving a counterexample, as in classical mathematics. However, it is also possible to give a **Brouwerian counterexample** to show that the statement is non-constructive. This sort of counterexample shows that the statement implies some principle that is known to be non-constructive. If it can be proved constructively that a statement implies some principle that is not constructively provable, then the statement itself cannot be constructively provable. For example, a particular statement may be shown to imply the law of the excluded middle. An example of a Brouwerian counterexample of this type is Diaconescu's theorem, which shows that the full axiom of choice is non-constructive in systems of constructive set theory, since the axiom of choice implies the law of excluded middle in such systems. The field of constructive reverse mathematics develops this idea further by classifying various principles in terms of "how nonconstructive" they are, by showing they are equivalent to various fragments of the law of the excluded middle.

Brouwer also provided "weak" counterexamples.[3] Such counterexamples do not disprove a statement, however; they only show that, at present, no constructive proof of the statement is known. One weak counterexample begins by taking some unsolved problem of mathematics, such as Goldbach's conjecture. Define a function f of a natural number x as follows:

$$f(x) = \begin{cases} 0 & \text{if Goldbach's conjecture is false} \\ 1 & \text{if Goldbach's conjecture is true} \end{cases}$$

Although this is a definition by cases, it is still an admissible definition in constructive mathematics. Several facts about f can be proved constructively. However, based on the different meaning of the words in constructive mathematics, if there is a constructive proof that "$f(0) = 1$ or $f(0) \neq 1$" then this would mean that there is a constructive proof of Goldbach's conjecture (in the former case) or a constructive proof that Goldbach's conjecture is false (in the latter case). Because no such proof is known, the quoted statement must also not have a known constructive proof. However, it is entirely possible that Goldbach's conjecture may have a constructive proof (as we do not know at present whether it does), in which case the quoted statement would have a constructive proof as well, albeit one that is unknown at present. The main practical use of weak counterexamples is to identify the "hardness" of a problem. For example, the counterexample just shown shows that the quoted statement is "at least as hard to prove" as Goldbach's conjecture. Weak counterexamples of this sort are often related to the limited principle of omniscience.

20.3   See also

- Existence theorem#'Pure' existence results

- Non-constructive algorithm existence proofs

- Errett Bishop - author of the book "Foundations of Constructive Analysis".

20.4   References

[1] J. Roger Hindley, "The Root-2 Proof as an Example of Non-constructivity", unpublished paper, September 2014, full text

[2] Dov Jarden, "A simple proof that a power of an irrational number to an irrational exponent may be rational", *Curiosa* No. 339 in *Scripta Mathematica* **19**:229 (1953)

[3] A. S. Troelstra, *Principles of Intuitionism*, Lecture Notes in Mathematics 95, 1969, p. 102

20.5   Further reading

- J. Franklin and A. Daoud (2011) *Proof in Mathematics: An Introduction.* Kew Books, ISBN 0-646-54509-4, ch. 4

- Hardy, G.H. & Wright, E.M. (1979) *An Introduction to the Theory of Numbers* (Fifth Edition). Oxford University Press. ISBN 0-19-853171-0

- Anne Sjerp Troelstra and Dirk van Dalen (1988) "Constructivism in Mathematics: Volume 1" Elsevier Science. ISBN 978-0-444-70506-8

20.6   External links

- *Weak counterexamples* by Mark van Atten, Stanford Encyclopedia of Philosophy

Chapter 21

Proof by exhaustion

This article is about the type of mathematical proof. For the method of calculating limits, see Method of exhaustion. "Brute force method" redirects here. For similarly named methods in other disciplines, see Brute force (disambiguation).

Proof by exhaustion, also known as **proof by cases**, **perfect induction**, or the **brute force method**, is a method of mathematical proof in which the statement to be proved is split into a finite number of cases or sets of equivalent cases and each type of case is checked to see if the proposition in question holds.[1] This is a method of direct proof. A proof by exhaustion contains two stages:

1. A proof that the cases are exhaustive; i.e., that each instance of the statement to be proved matches the conditions of (at least) one of the cases.

2. A proof of each of the cases.

In the Curry–Howard isomorphism, proof by exhaustion and case analysis are related to ML-style pattern matching.

21.1   Example

To prove that every integer that is a perfect cube is a multiple of 9, or is 1 more than a multiple of 9, or is 1 less than a multiple of 9.

Proof:
Each cube number is the cube of some integer n. Every integer n is either a multiple of 3, or 1 more or 1 less than a multiple of 3. So these 3 cases are exhaustive:

- Case 1: If $n = 3p$, then $n^3 = 27p^3$, which is a multiple of 9.
- Case 2: If $n = 3p + 1$, then $n^3 = 27p^3 + 27p^2 + 9p + 1$, which is 1 more than a multiple of 9. For instance, if $n = 4$ then $n^3 = 64 = 9 \times 7 + 1$.
- Case 3: If $n = 3p - 1$, then $n^3 = 27p^3 - 27p^2 + 9p - 1$, which is 1 less than a multiple of 9. For instance, if $n = 5$ then $n^3 = 125 = 9 \times 14 - 1.$∎

21.2   Number of cases

There is no upper limit to the number of cases allowed in a proof by exhaustion. Sometimes there are only two or three cases. Sometimes there may be thousands or even millions. For example, rigorously solving an endgame puzzle in chess might involve considering a very large number of possible positions in the game tree of that problem.

The first proof of the four colour theorem was a proof by exhaustion with 1,936 cases. This proof was controversial because the majority of the cases were checked by a computer program, not by hand. The shortest known proof of the four colour theorem today still has over 600 cases.

Mathematicians prefer to avoid proofs with large numbers of cases, as they seem inelegant, and in general the probability of an error in the whole proof increases with the number of cases. A proof with a large number of cases leaves an impression that the theorem is only true by coincidence, and not because of some underlying principle or connection. Other types of proofs—such as proof by induction (mathematical induction)—are considered more elegant. However, there are some important theorems for which no other method of proof has been found, such as

- The proof that there is no finite projective plane of order 10.

- The classification of finite simple groups.

- The Kepler conjecture.

21.3   See also

- Case analysis

- Computer-assisted proof

21.4   Notes

[1] Reid, D. A; Knipping, C (2010), *Proof in Mathematics Education: Research, Learning, and Teaching*, Sense Publishers, p. 133, ISBN 978-9460912443.

Chapter 22

Probabilistic method

This article is not about interactive proof systems which use probability to convince a verifier that a proof is correct, nor about probabilistic algorithms, which give the right answer with high probability but not with certainty, nor about Monte Carlo methods, which are simulations relying on pseudo-randomness.

The **probabilistic method** is a nonconstructive method, primarily used in combinatorics and pioneered by Paul Erdős, for proving the existence of a prescribed kind of mathematical object. It works by showing that if one randomly chooses objects from a specified class, the probability that the result is of the prescribed kind is more than zero. Although the proof uses probability, the final conclusion is determined for *certain*, without any possible error.

This method has now been applied to other areas of mathematics such as number theory, linear algebra, and real analysis, as well as in computer science (e.g. randomized rounding), and information theory.

22.1 Introduction

If every object in a collection of objects fails to have a certain property, then the probability that a random object chosen from the collection has that property is zero. Turning this around, if the probability that the random object has the property is greater than zero, then this proves the existence of at least one object in the collection that has the property. It doesn't matter if the probability is vanishingly small; any positive probability will do.

Similarly, showing that the probability is (strictly) less than 1 can be used to prove the existence of an object that does *not* satisfy the prescribed properties.

Another way to use the probabilistic method is by calculating the expected value of some random variable. If it can be shown that the random variable can take on a value less than the expected value, this proves that the random variable can also take on some value greater than the expected value.

Common tools used in the probabilistic method include Markov's inequality, the Chernoff bound, and the Lovász local lemma.

22.2 Two examples due to Erdős

Although others before him proved theorems via the probabilistic method (for example, Szele's 1943 result that there exist tournaments containing a large number of Hamiltonian cycles), many of the most well known proofs using this method are due to Erdős. Indeed, the Alon-Spencer textbook on the subject has his picture on the cover to highlight the method's association with Erdős. The first example below describes one such result from 1947 that gives a proof of a lower bound for the Ramsey number $R(r, r)$.

22.2.1   First example

Suppose we have a complete graph on n vertices. We wish to show (for small enough values of n) that it is possible to color the edges of the graph in two colors (say red and blue) so that there is no complete subgraph on r vertices which is monochromatic (every edge colored the same color).

To do so, we color the graph randomly. Color each edge independently with probability 1/2 of being red and 1/2 of being blue. We calculate the expected number of monochromatic subgraphs on r vertices as follows:

For any set S of r vertices from our graph, define the variable $X(S)$ to be 1 if every edge amongst the r vertices is the same color, and 0 otherwise. Note that the number of monochromatic r-subgraphs is the sum of $X(S)$ over all possible subsets. For any S, the expected value of $X(S)$ is simply the probability that all of the

$$\binom{r}{2}$$

edges in S are the same color,

$$2 \cdot 2^{-\binom{r}{2}}$$

(the factor of 2 comes because there are two possible colors).

This holds true for any of the $C(n, r)$ possible subsets we could have chosen, so we have that the sum of $E[X(S)]$ over all S is

$$\binom{n}{r} 2^{1-\binom{r}{2}}.$$

The sum of an expectation is the expectation of the sum (*regardless* of whether the variables are independent), so the expectation of the sum (the expected number of monochromatic r-subgraphs) is

$$\binom{n}{r} 2^{1-\binom{r}{2}}.$$

Consider what happens if this value is less than 1. The number of monochromatic r-subgraphs in our random coloring will always be an integer, so at least one coloring must have less than the expected value. But the only integer that satisfies this criterion is 0. Thus if

$$\binom{n}{r} < 2^{\binom{r}{2}-1},$$

(which holds, for example, for n=5 and r=4) then some coloring fits our desired criterion.[1]

By definition of the Ramsey number, this implies that $R(r, r)$ must be bigger than n. In particular, $R(r, r)$ must grow at least exponentially with r.

A peculiarity of this argument is that it is entirely nonconstructive. Even though it proves (for example) that almost every coloring of the complete graph on $(1.1)^r$ vertices contains no monochromatic r-subgraph, it gives no explicit example of such a coloring. The problem of finding such a coloring has been open for more than 50 years.

22.2.2   Second example

A 1959 paper of Erdős (see reference cited below) addressed the following problem in graph theory: given positive integers g and k, does there exist a graph G containing only cycles of length at least g, such that the chromatic number of G is at least k?

It can be shown that such a graph exists for any g and k, and the proof is reasonably simple. Let n be very large and consider a random graph G on n vertices, where every edge in G exists with probability $p = n^{1/g-1}$. It can be shown that with positive probability, the following two properties hold:

Property 1. G contains at most $n/2$ cycles of length less than g.

Proof. Let X be the number cycles of length less than g. Number of cycles of length i in the complete graph on n vertices is

$$\frac{n!}{2 \cdot i \cdot (n-i)!} \leq \frac{n^i}{2}$$

and each of them is present in G with probability p^i. Hence by Markov's inequality we have

$$\Pr\left(X > \tfrac{n}{2}\right) \leq \frac{2}{n} E[X] \leq \frac{1}{n} \sum_{i=3}^{g-1} p^i n^i = \frac{1}{n} \sum_{i=3}^{g-1} n^{\frac{i}{g}} \leq \frac{g}{n} n^{\frac{g-1}{g}} \leq g n^{-\frac{1}{g}} = o(1).$$

Property 2. G contains no independent set of size $\lceil \frac{n}{2k} \rceil$.

Proof. Let Y be the size of the largest independent set in G. Clearly, we have

$$\Pr(Y \geq y) \leq \binom{n}{y} (1-p)^{\frac{y(y-1)}{2}} \leq n^y e^{-\frac{py(y-1)}{2}} = e^{-\frac{y}{2} \cdot (py - 2 \ln n - p)} = o(1),$$

when

$$y = \left\lceil \frac{n}{2k} \right\rceil.$$

Here comes the trick: since G has these two properties, we can remove at most $n/2$ vertices from G to obtain a new graph G' on n' vertices that contains only cycles of length at least g. We can see that this new graph has no independent set of size $\lceil \frac{n'}{k} \rceil$. Hence G' has chromatic number at least k, as chromatic number is lower bounded by 'number of vertices/size of largest independent set'.

This result gives a hint as to why the computation of the chromatic number of a graph is so difficult: even when there are no local reasons (such as small cycles) for a graph to require many colors the chromatic number can still be arbitrarily large.

22.3   See also

- Random graph

- Probabilistic proofs of non-probabilistic theorems

- Method of conditional probabilities

- Interactive proof system

22.4   References

- Alon, Noga; Spencer, Joel H. (2000). *The probabilistic method* (2ed). New York: Wiley-Interscience. ISBN 0-471-37046-0.

- Erdős, P. (1959). "Graph theory and probability" (PDF). *Canad. J. Math.* **11** (0): 34–38. doi:10.4153/CJM-1959-003-9. MR 0102081.

- Erdős, P. (1961). "Graph theory and probability, II" (PDF). *Canad. J. Math.* **13** (0): 346–352. doi:10.4153/CJM-1961-029-9. MR 0120168.

- J. Matoušek, J. Vondrak. The Probabilistic Method. Lecture notes.

- Alon, N and Krivelevich, M (2006). Extremal and Probabilistic Combinatorics

22.5   Footnotes

[1] The same fact can be proved without probability, using a simple counting argument:

- The total number of r-subgraphs is $\binom{n}{r}$.

- Each r-subgraphs has $\binom{r}{2}$ edges and thus can be colored in $2^{\binom{r}{2}}$ different ways.

- Of these colorings, only 2 colorings are 'bad' for that subgraph (the colorings in which all vertices are red or all vertices are blue).

- Hence, the total number of colorings that are bad for *all* subgraphs is at most $2\binom{n}{r}$.

- Hence, if $2^{\binom{r}{2}} > 2\binom{n}{r}$, there must be at least one coloring which is not 'bad' for any subgraph.

Chapter 23

Combinatorial proof

In mathematics, the term ***combinatorial proof*** is often used to mean either of two types of mathematical proof:

- A proof by double counting. A combinatorial identity is proven by counting the number of elements of some carefully chosen set in two different ways to obtain the different expressions in the identity. Since those expressions count the same objects, they must be equal to each other and thus the identity is established.

- A bijective proof. Two sets are shown to have the same number of members by exhibiting a bijection, i.e. a one-to-one correspondence, between them.

The term "combinatorial proof" may also be used more broadly to refer to any kind of elementary proof in combinatorics. However, as Glass (2003) writes in his review of Benjamin & Quinn (2003) (a book about combinatorial proofs), these two simple techniques are enough to prove many theorems in combinatorics and number theory.

23.1 Example

An archetypal double counting proof is for the well known formula for the number $\binom{n}{k}$ of k-combinations (i.e., subsets of size k) of an n-element set:

$$\binom{n}{k} = \frac{n(n-1)\cdots(n-k+1)}{k(k-1)\cdots 1}.$$

Here a direct bijective proof is not possible: because the right-hand side of the identity is a fraction, there is no set *obviously* counted by it (it even takes some thought to see that the denominator always evenly divides the numerator). However its numerator counts the Cartesian product of k finite sets of sizes n, $n-1$, ..., $n-k+1$, while its denominator counts the permutations of a k-element set (the set most obviously counted by the denominator would be another Cartesian product k finite sets; if desired one could map permutations to that set by an explicit bijection). Now take S to be the set of sequences of k elements selected from our n-element set without repetition. On one hand, there is an easy bijection of S with the Cartesian product corresponding to the numerator $n(n-1)\cdots(n-k+1)$, and on the other hand there is a bijection from the set C of pairs of a k-combination and a permutation σ of k to S, by taking the elements of C in increasing order, and then permuting this sequence by σ to obtain an element of S. The two ways of counting give the equation

$$n(n-1)\cdots(n-k+1) = \binom{n}{k}k!,$$

and after division by $k!$ this leads to the stated formula for $\binom{n}{k}$. In general, if the counting formula involves a division, a similar double counting argument (if it exists) gives the most straightforward combinatorial proof of the identity, but double counting arguments are not limited to situations where the formula is of this form.

23.2   The benefit of a combinatorial proof

Stanley (1997) gives an example of a combinatorial enumeration problem (counting the number of sequences of k subsets $S_1, S_2, \ldots Sk$, that can be formed from a set of n items such that the subsets have an empty common intersection) with two different proofs for its solution. The first proof, which is not combinatorial, uses mathematical induction and generating functions to find that the number of sequences of this type is $(2^k - 1)^n$. The second proof is based on the observation that there are $2^k - 1$ proper subsets of the set $\{1, 2, \ldots, k\}$, and $(2^k - 1)^n$ functions from the set $\{1, 2, \ldots, n\}$ to the family of proper subsets of $\{1, 2, \ldots, k\}$. The sequences to be counted can be placed in one-to-one correspondence with these functions, where the function formed from a given sequence of subsets maps each element i to the set $\{j \mid i \in Sj\}$.

Stanley writes, "Not only is the above combinatorial proof much shorter than our previous proof, but also it makes the reason for the simple answer completely transparent. It is often the case, as occurred here, that the first proof to come to mind turns out to be laborious and inelegant, but that the final answer suggests a simple combinatorial proof." Due both to their frequent greater elegance than non-combinatorial proofs and the greater insight they provide into the structures they describe, Stanley formulates a general principle that combinatorial proofs are to be preferred over other proofs, and lists as exercises many problems of finding combinatorial proofs for mathematical facts known to be true through other means.

23.3   The difference between bijective and double counting proofs

Stanley does not clearly distinguish between bijective and double counting proofs, and gives examples of both kinds, but the difference between the two types of combinatorial proof can be seen in an example provided by Aigner & Ziegler (1998), of proofs for Cayley's formula stating that there are n^{n-2} different trees that can be formed from a given set of n nodes. Aigner and Ziegler list four proofs of this theorem, the first of which is bijective and the last of which is a double counting argument. They also mention but do not describe the details of a fifth bijective proof.

The most natural way to find a bijective proof of this formula would be to find a bijection between n-node trees and some collection of objects that has n^{n-2} members, such as the sequences of $n-2$ values each in the range from 1 to n. Such a bijection can be obtained using the Prüfer sequence of each tree. Any tree can be uniquely encoded into a Prüfer sequence, and any Prüfer sequence can be uniquely decoded into a tree; these two results together provide a bijective proof of Cayley's formula.

An alternative bijective proof, given by Aigner and Ziegler and credited by them to André Joyal, involves a bijection between, on the one hand, n-node trees with two designated nodes (that may be the same as each other), and on the other hand, n-node directed pseudoforests. If there are Tn n-node trees, then there are $n^2 Tn$ trees with two designated nodes. And a pseudoforest may be determined by specifying, for each of its nodes, the endpoint of the edge extending outwards from that node; there are n possible choices for the endpoint of a single edge (allowing self-loops) and therefore n^n possible pseudoforests. By finding a bijection between trees with two labeled nodes and pseudoforests, Joyal's proof shows that $Tn = n^{n-2}$.

Finally, the fourth proof of Cayley's formula presented by Aigner and Ziegler is a double counting proof due to Jim Pitman. In this proof, Pitman considers the sequences of directed edges that may be added to an n-node empty graph to form from it a single rooted tree, and counts the number of such sequences in two different ways. By showing how to derive a sequence of this type by choosing a tree, a root for the tree, and an ordering for the edges in the tree, he shows that there are $Tnn!$ possible sequences of this type. And by counting the number of ways in which a partial sequence can be extended by a single edge, he shows that there are $n^{n-2}n!$ possible sequences. Equating these two different formulas for the size of the same set of edge sequences and cancelling the common factor of $n!$ leads to Cayley's formula.

23.4 Related concepts

- The principles of double counting and bijection used in combinatorial proofs can be seen as examples of a larger family of combinatorial principles, which include also other ideas such as the pigeonhole principle.

- Proving an identity combinatorially can be viewed as adding more structure to the identity by replacing numbers by sets; similarly, categorification is the replacement of sets by categories.

23.5 References

- Aigner, Martin; Ziegler, Günter M. (1998), *Proofs from THE BOOK*, Springer-Verlag, pp. 141–146, ISBN 3-540-40460-0.

- Benjamin, Arthur T.; Quinn, Jennifer J. (2003), *Proofs that Really Count: The Art of Combinatorial Proof*, Dolciani Mathematical Expositions **27**, Mathematical Association of America, ISBN 978-0-88385-333-7.

- Glass, Darren (2003), *Read This: Proofs that Really Count*, Mathematical Association of America.

- Stanley, Richard P. (1997), *Enumerative Combinatorics, Volume I*, Cambridge Studies in Advanced Mathematics **49**, Cambridge University Press, pp. 11–12, ISBN 0-521-55309-1.

Chapter 24

Statistical proof

Statistical proof is the rational demonstration of degree of certainty for a proposition, hypothesis or theory that is used to convince others subsequent to a statistical test of the supporting evidence and the types of inferences that can be drawn from the test scores. Statistical methods are used to increase the understanding of the facts and the proof demonstrates the validity and logic of inference with explicit reference to a hypothesis, the experimental data, the facts, the test, and the odds. Proof has two essential aims: the first is to convince and the second is to explain the proposition through peer and public review.[1]

The burden of proof rests on the demonstrable application of the statistical method, the disclosure of the assumptions, and the relevance that the test has with respect to a genuine understanding of the data relative to the external world. There are adherents to several different statistical philosophies of inference, such as Bayes theorem versus the likelihood function, or positivism versus critical rationalism. These methods of reason have direct bearing on statistical proof and its interpretations in the broader philosophy of science.[1][2]

A common demarcation between science and non-science is the hypothetico-deductive proof of falsification developed by Karl Popper, which is a well-established practice in the tradition of statistics. Other modes of inference, however, may include the inductive and abductive modes of proof.[3] Scientists do not use statistical proof as a means to attain certainty, but to falsify claims and explain theory. Science cannot achieve absolute certainty nor is it a continuous march toward an objective truth as the vernacular as opposed to the scientific meaning of the term "proof" might imply. Statistical proof offers a kind of proof of a theory's falsity and the means to learn heuristically through repeated statistical trials and experimental error.[2] Statistical proof also has applications in legal matters with implications for the legal burden of proof.[4]

24.1 Axioms

There are two kinds of axioms, 1) conventions that are taken as true that should be avoided because they cannot be tested, and 2) hypotheses.[5] Proof in the theory of probability was built on four axioms developed in the late 17th century:

1. The probability of a hypotheses is a non-negative real number: $\left\{ \Pr(h) \geqq 0 \right\}$;

2. The probability of necessary truth equals one: $\left\{ \Pr(t) = 1 \right\}$;

3. If two hypotheses h_1 and h_2 are mutually exclusive, then the sum of their probabilities is equal to the probability of their disjunction: $\left\{ \Pr(h_1) + \Pr(h_2) = \Pr(h_1 \, or \, h_2) \right\}$;

4. The conditional probability of h_1 given h_2 $\left\{ \Pr(h_1|h_2) \right\}$ is equal to the unconditional probability $\left\{ \Pr(h_1 \quad \& \right.$

144

h_2} of the conjunction h_1 and h_2, divided by the unconditional probability { $\Pr(h_2)$ } of h_2 where that probability is positive { $\frac{\Pr(h_1|h_2)=\Pr(h_1 \, \& \, h_2)}{\Pr(h_2)}$ } , where { $\Pr(h_2) > 0$ } .

The preceding axioms provide the statistical proof and basis for the laws of randomness, or objective chance from where modern statistical theory has advanced. Experimental data, however, can never prove that the hypotheses (h) is true, but relies on an inductive inference by measuring the probability of the hypotheses relative to the empirical data. The proof is in the rational demonstration of using the logic of inference, math, testing, and deductive reasoning of significance.[1][2][6]

24.2 Test and proof

Main article: Statistical tests

The term *proof* descended from its Latin roots (provable, probable, *probare* L.) meaning *to test*.[7][8] Hence, proof is a form of inference by means of a statistical test. Statistical tests are formulated on models that generate probability distributions. Examples of probability distributions might include the binary, normal, or poisson distribution that give exact descriptions of variables that behave according to natural laws of random chance. When a statistical test is applied to samples of a population, the test determines if the sample statistics are significantly different from the assumed null-model. True values of a population, which are unknowable in practice, are called parameters of the population. Researchers sample from populations, which provide estimates of the parameters, to calculate the mean or standard deviation. If the entire population is sampled, then the sample statistic mean and distribution will converge with the parametric distribution.[9]

Using the scientific method of falsification, the probability value that the sample statistic is sufficiently different from the null-model than can be explained by chance alone is given prior to the test. Most statisticians set the prior probability value at 0.05 or 0.1, which means if the sample statistics diverge from the parametric model more than 5 (or 10) times out of 100, then the discrepancy is unlikely to be explained by chance alone and the null-hypothesis is rejected. Statistical models provide exact outcomes of the parametric and estimates of the sample statistics. Hence, the burden of proof rests in the sample statistics that provide estimates of a statistical model. Statistical models contain the mathematical proof of the parametric values and their probability distributions.[10][11]

24.3 Bayes theorem

Main article: Bayes theorem
See also: Evidence under Bayes theorem

Bayesian statistics are based on a different philosophical approach for proof of inference. The mathematical formula for Bayes's theorem is:

$Pr[Parameter|Data] = \frac{Pr[Data|Parameter] \times Pr[Parameter]}{Pr[Data]}$

The formula is read as the probability of the parameter (or hypothesis =h, as used in the notation on axioms) "given" the data (or empirical observation), where the horizontal bar refers to "given". The right hand side of the formula calculates the prior probability of a statistical model (Pr [Parameter]) with the likelihood (Pr [Data | Parameter]) to produce a posterior probability distribution of the parameter (Pr [Parameter | Data]). The posterior probability is the likelihood that the parameter is correct given the observed data or samples statistics.[12] Hypotheses can be compared using Bayesian inference by means of the Bayes factor, which is the ratio of the posterior odds to the prior odds. It provides a measure of the data and if it has increased or decreased the likelihood of one hypotheses relative to another.[13]

The statistical proof is the Bayesian demonstration that one hypothesis has a higher (weak, strong, positive) likelihood.[13] There is considerable debate if the Bayesian method aligns with Karl Poppers method of proof of falsification, where some have suggested that "...there is no such thing as "accepting" hypotheses at all. All that one does in science is assign degrees of belief..."[14]:180 According to Popper, hypotheses that have withstood testing and have yet to be falsified are not verified but corroborated. Some researches have suggested that Popper's quest to define corroboration on the premise of probability put his philosophy in line with the Bayesian approach. In this context, the likelihood of one hypothesis relative to another may be an index of corroboration, not confirmation, and thus statistically proven through rigorous objective standing.[6][15]

24.4 In legal proceedings

Main article: Legal burden of proof

"Where gross statistical disparities can be shown, they alone may in a proper case constitute *prima facie* proof of a pattern or practice of discrimination."[nb 1]:271

Statistical proof in a legal proceeding can be sorted into three categories of evidence:

1. The occurrence of an event, act, or type of conduct,

2. The identity the individual(s) responsible

3. The intent or psychological responsibility[16]

Statistical proof was not regularly applied in decisions concerning United States legal proceedings until the mid 1970's following a landmark jury discrimination case in *Castaneda v. Partida*. The US Supreme Court ruled that gross statistical disparities constitutes *"prima facie* proof" of discrimination, resulting in a shift of the burden of proof from plaintiff to defendant. Since that ruling, statistical proof has been used in many other cases on inequality, discrimination, and DNA evidence.[4][17][18] However, there is not a one-to-one correspondence between statistical proof and the legal burden of proof. "The Supreme Court has stated that the degrees of rigor required in the fact finding processes of law and science do not necessarily correspond."[18]:1533

In an example of a death row sentence (*McCleskey v. Kemp*[nb 2]) concerning racial discrimination, the petitioner, a black man named McCleskey was charged with the murder of a white police officer during a robbery. Expert testimony for McClesky introduced a statistical proof showing that "defendants charged with killing white victims were 4.3 times as likely to receive a death sentence as charged with killing blacks.".[19]:595 Nonetheless, the statistics was insufficient "to prove that the decisionmakers in his case acted with discriminatory purpose."[19]:596 It was further argued that there were "inherent limitations of the statistical proof",[19]:596 because it did not refer to the specifics of the individual. Despite the statistical demonstration of an increased probability of discrimination, the legal burden of proof (it was argued) had to be examined on a case by case basis.[19]

24.5 See also

- Mathematical proof

- Data analysis

24.6 References

[1] Gold, B.; Simons, R. A. (2008). *Proof and other dilemmas: Mathematics and philosophy*. Mathematics Association of America Inc. ISBN 0-88385-567-4.

[2] Gattei, S. (2008). *Thomas Kuhn's "Linguistic Turn" and the Legacy of Logical Empiricism: Incommensurability, Rationality and the Search for Truth*. Ashgate Pub Co. p. 277. ISBN 0-7546-6160-1.

[3] Pedemont, B. (2007). "How can the relationship between argumentation and proof be analysed?". *Educational Studies in Mathematics* **66** (1): 23–41. doi:10.1007/s10649-006-9057-x.

[4] Meier, P. (1986). "Damned Liars and Expert Witnesses" (PDF). *Journal of the American Statistical Association* **81** (394): 269–276. doi:10.1080/01621459.1986.10478270.

[5] Wiley, E. O. (1974). "Karl R. Popper, Systematics, and Classification: A Reply to Walter Bock and Other Evolutionary Taxonomists" (PDF). *Systematic Biology* **24** (2): 233–243. doi:10.1093/sysbio/24.2.233.

[6] Howson, C.; Urbach, P. (1991). "Bayesian reasoning in science" (PDF). *Nature* **350** (6317): 371–374. doi:10.1038/350371a0.

[7] Sundholm, G. "Proof-Theoretical Semantics and Fregean Identity Criteria for Propositions" (PDF). *The Monist* **77** (3): 294–314. doi:10.5840/monist199477315.

[8] Bissell, D. (1996). "Statisticians have a Word for it" (PDF). *Teaching Statistics* **18** (3): 87–89. doi:10.1111/j.1467-9639.1996.tb00300.x

[9] Sokal, R. R.; Rohlf, F. J. (1995). *Biometry* (3rd ed.). W.H. Freeman & Company. p. 887. ISBN 0-7167-2411-1.

[10] Heath, David (1995). *An introduction to experimental design and statistics for biology*. CRC Press. ISBN 1-85728-132-2.

[11] Hald, Anders (2006). *A History of Parametric Statistical Inference from Bernoulli to Fisher, 1713-1935*. Springer. p. 260. ISBN 0-387-46408-5.

[12] Huelsenbeck, J. P.; Ronquist, F.; Bollback, J. P. (2001). "Bayesian Inference of Phylogeny and Its Impact on Evolutionary Biology" (PDF). *Science* **294** (5550): 2310–2314. doi:10.1126/science.1065889.

[13] Wade, P. R. (2000). "Bayesian methods in conservation biology" (PDF). *Conservation Biology* **14** (5): 1308–1316. doi:10.1046/j.1523-1739.2000.99415.x.

[14] Sober, E. (1991). *Reconstructing the Past: Parsimony, Evolution, and Inference*. A Bradford Book. p. 284. ISBN 0-262-69144-2.

[15] Helfenbein, K. G.; DeSalle, R. (2005). "Falsifications and corroborations: Karl Popper's influence on systematics" (PDF). *Molecular Phylogenetics and Evolution* **35**: 271–280. doi:10.1016/j.ympev.2005.01.003.

[16] Fienberg, S. E.; Kadane, J. B. "The presentation of Bayesian statistical analyses in legal proceedings". *Journal of the Royal Statistical Society, Series D* **32** (1/2): 88–98. doi:10.2307/2987595. JSTOR 2987595.

[17] Garaud, M. C. (1990). "Legal Standards and Statistical Proof in Title VII Litigation: In Search of a Coherent Disparate Impact Model". *University of Pennsylvania Law Review* **139** (2): 455–503. JSTOR 3312286.

[18] The Harvard Law Review Association (1995). "Developments in the Law: Confronting the New Challenges of Scientific Evidence". *Harvard Law Review* **108** (7): 1481–1605. doi:10.2307/1341808. JSTOR 1341808.

[19] Faigman, D. L. (1991). "Normative Constitutional Fact-Finding": Exploring the Empirical Component of Constitutional Interpretation". *University of Pennsylvania Law Review* **139** (3): 541–613. JSTOR 3312337.

24.7 Notes

[1] Supreme Court of the United States *Castaneda v. Partida*, 1977 cited in Meier (1986) Ibid. who states "Thus, in the space of less than half a year, the Supreme Court had moved from the traditional legal disdain for statistical proof to a strong endorsement of it as being capable, on its own, of establishing a prima facie case against a defendant."[4]

[2] 481 U.S. 279 (1987).[19]

Chapter 25

Computer-assisted proof

A **computer-assisted proof** is a mathematical proof that has been at least partially generated by computer.

Most computer-aided proofs to date have been implementations of large proofs-by-exhaustion of a mathematical theorem. The idea is to use a computer program to perform lengthy computations, and to provide a proof that the result of these computations implies the given theorem. In 1976, the four color theorem was the first major theorem to be verified using a computer program.

Attempts have also been made in the area of artificial intelligence research to create smaller, explicit, new proofs of mathematical theorems from the bottom up using machine reasoning techniques such as heuristic search. Such automated theorem provers have proved a number of new results and found new proofs for known theorems. Additionally, interactive proof assistants allow mathematicians to develop human-readable proofs which are nonetheless formally verified for correctness. Since these proofs are generally human-surveyable (albeit with difficulty, as with the proof of the Robbins conjecture) they do not share the controversial implications of computer-aided proofs-by-exhaustion.

25.1 Methods

One method for using computers in mathematical proofs is by means of so-called validated numerics or rigorous numerics. This means computing numerically yet with mathematical rigour. One uses set-valued arithmetic and inclusion principle in order to ensure that the set-valued output of a numerical program encloses the solution of the original mathematical problem. This is done by controlling, enclosing and propagating round-off and truncation errors using for example interval arithmetic. More precisely, one reduces the computation to a sequence of elementary operations, say (+,-,*,/). In a computer, the result of each elementary operation is rounded off by the computer precision. However, one can construct an interval provided by upper and lower bounds on the result of an elementary operation. Then one proceeds by replacing numbers with intervals and performing elementary operations between such intervals of representable numbers.

25.2 Philosophical objections

Computer-assisted proofs are the subject of some controversy in the mathematical world. Some mathematicians believe that lengthy computer-assisted proofs are not, in some sense, 'real' mathematical proofs because they involve so many logical steps that they are not practically verifiable by human beings, and that mathematicians are effectively being asked to replace logical deduction from assumed axioms with trust in an empirical computational process, which is potentially affected by errors in the computer program, as well as defects in the runtime environment and hardware.

Other mathematicians believe that lengthy computer-assisted proofs should be regarded as *calculations*, rather than *proofs*: the proof algorithm itself should be proved valid, so that its use can then be regarded as a mere "verification". Arguments that computer-assisted proofs are subject to errors in their source programs, compilers, and hardware can be resolved by providing a formal proof of correctness for the computer program (an approach which was successfully applied to the four-color theorem in 2005) as well as replicating the result using different programming languages, different compilers, and different computer hardware.

Another possible way of verifying computer-aided proofs is to generate their reasoning steps in a machine-readable form, and then use an automated theorem prover to demonstrate their correctness. This approach of using a computer program to prove another program correct does not appeal to computer proof skeptics, who see it as adding another layer of complexity without addressing the perceived need for human understanding.

Another argument against computer-aided proofs is that they lack mathematical elegance—that they provide no insights or new and useful concepts. In fact, this is an argument that could be advanced against any lengthy proof by exhaustion.

An additional philosophical issue raised by computer-aided proofs is whether they make mathematics into a quasi-empirical science, where the scientific method becomes more important than the application of pure reason in the area of abstract mathematical concepts. This directly relates to the argument within mathematics as to whether mathematics is based on ideas, or "merely" an exercise in formal symbol manipulation. It also raises the question whether, if according to the Platonist view, all possible mathematical objects in some sense "already exist", whether computer-aided mathematics is an observational science like astronomy, rather than an experimental one like physics or chemistry. Interestingly, this controversy within mathematics is occurring at the same time as questions are being asked in the physics community about whether twenty-first century theoretical physics is becoming too mathematical, and leaving behind its experimental roots.

The emerging field of experimental mathematics is confronting this debate head-on by focusing on numerical experiments as its main tool for mathematical exploration.

25.3 Theorems for sale

In 2010, academics at The University of Edinburgh offered people the chance to "buy their own theorem" created through a computer-assisted proof. This new theorem would be named after the purchaser.[1]

25.4 List of theorems proved with the help of computer programs

Inclusion in this list does not imply that a formal computer-checked proof exists, but rather, that a computer program has been involved in some way. See the main articles for details.

- Four color theorem, 1976

- Mitchell Feigenbaum's universality conjecture in non-linear dynamics. Proven by O. E. Lanford using rigorous computer arithmetic, 1982

- Connect Four, 1988 – a solved game

- Non-existence of a finite projective plane of order 10, 1989

- Robbins conjecture, 1996

- Kepler conjecture, 1998 – the problem of optimal sphere packing in a box

- Lorenz attractor, 2002 – 14th of Smale's problems proved by W. Tucker using interval arithmetic

- 17-point case of the Happy Ending problem, 2006

- NP-hardness of minimum-weight triangulation, 2008

- Optimal solutions for Rubik's Cube can be obtained in at most 20 face moves, 2010

- Minimum number of clues for a solvable Sudoku puzzle is 17, 2012

25.5   See also

- Mathematical proof
- Model checking
- Proof checking
- Symbolic computation
- Automated reasoning
- Formal verification
- Seventeen or Bust
- Metamath

25.6   References

[1] "Herald Gazette article on buying your own theorem". *Herald Gazette Scotland*. November 2010.

25.7   Further reading

- Lenat, D.B., (1976), AM: An artificial intelligence approach to discovery in mathematics as heuristic search, Ph.D. Thesis, STAN-CS-76-570, and Heuristic Programming Project Report HPP-76-8, Stanford University, AI Lab., Stanford, CA.

25.8   External links

- Oscar E. Lanford; A computer-assisted proof of the Feigenbaum conjectures, "Bull. Amer. Math. Soc.", 1982
- Edmund Furse; Why did AM run out of steam?
- Number proofs done by computer might err
- "A Special Issue on Formal Proof". *Notices of the American Mathematical Society*. December 2008.

Chapter 26

Psychologism

Psychologism is a philosophical position, according to which psychology plays a central role in grounding or explaining some other, non-psychological type of fact or law. The *Oxford English Dictionary* defines *psychologism* as: "The view or doctrine that a theory of psychology or ideas forms the basis of an account of metaphysics, epistemology, or meaning; (sometimes) spec. the explanation or derivation of mathematical or logical laws in terms of psychological facts."[1] Psychologism in epistemology, the idea that its problems "can be solved satisfactorily by the psychological study of the development of mental processes", was argued in John Locke's *An Essay Concerning Human Understanding* (1690).[2]

Other forms of psychologism are logical psychologism and mathematical psychologism. Logical psychologism is a position in logic (or the philosophy of logic) according to which logical laws and mathematical laws are grounded in, derived from, explained or exhausted by psychological facts (or laws). Psychologism in the philosophy of mathematics is the position that mathematical concepts and/or truths are grounded in, derived from or explained by psychological facts (or laws).

The word was coined by Johann Eduard Erdmann as *Psychologismus*, being translated into English as *psychologism*.

John Stuart Mill was accused by Edmund Husserl of being an advocate of a type of logical psychologism, although this may not have been the case.[3] So were many nineteenth-century German logicians such as Christoph von Sigwart and Erdmann himself, as well as a number of psychologists, past and present (e.g., Gustave Le Bon). Psychologism was criticized by Gottlob Frege in his *The Foundations of Arithmetic*, and many of his works and essays, including his review of Husserl's *Philosophy of Arithmetic*. Husserl, in the first volume of his *Logical Investigations*, called "The Prolegomena of Pure Logic", criticized psychologism thoroughly and sought to distance himself from it. Frege's arguments were largely ignored, while Husserl's were widely discussed.[4]

In "Psychologism and Behaviorism", Ned Block takes psychologism as the position that "whether behavior is intelligent behavior depends on the character of the internal information processing that produces it." This is in contrast to a behavioral view which would state that intelligence can be ascribed to a being solely via observing its behavior. This type of behavioral view is strongly associated with the Turing test.[5]

26.1 See also

- Anti-psychologism
- Naturalized epistemology
- Blockhead

26.2 References

[1] *Oxford English Dictionary*.

[2] *Encyclopedia Britannica*: Psychologism.

[3] Skorupski, JM., *Mill (Arguments of the Philosophers)*, Routledge, 2010, pp. 127-128.

[4] Kusch, Martin (Nov 7, 2011). "Psychologism". The Stanford Encyclopedia of Philosophy. Retrieved Nov 19, 2014.

[5] Block, Ned (1981), "Psychologism and Behaviorism", *The Philosophical Review* (Duke University Press) **90** (1): 5–43, doi:10.2307/2184371, JSTOR 2184371

26.3   External links

- Husserl's Criticism of Psychologism. Link broken, page preserved most recently from October 22, 2009 at Internet Archive: Eprint. From *Diwatao*, (apparently former) online journal of the philosophy department of San Beda College, Manila, the Philippines.

- The Turing test entry in the *Stanford Encyclopedia of Philosophy*

Chapter 27

Language of thought hypothesis

In philosophy of mind, the **language of thought hypothesis** (**LOTH**) put forward by American philosopher Jerry Fodor describes thoughts as represented in a "language" (sometimes known as *mentalese*) that allows complex thoughts to be built up by combining simpler thoughts in various ways. In its most basic form, the theory states that thought follows the same rules as language: thought has syntax.

Using empirical data drawn from linguistics and cognitive science to describe mental representation from a philosophical vantage-point, the hypothesis states that thinking takes place in a language of thought (LOT): cognition and cognitive processes are only 'remotely plausible' when expressed as a system of representations that is "tokened" by a linguistic or semantic structure and operated upon by means of a combinatorial syntax.[1] Linguistic tokens used in mental language describe elementary concepts which are operated upon by logical rules establishing causal connections to allow for complex thought. Syntax as well as semantics have a causal effect on the properties of this system of mental representations.

These mental representations are not present in the brain in the same way as symbols are present on paper; rather, the LOT is supposed to exist at the cognitive level, the level of thoughts and concepts. The LOTH has wide-ranging significance for a number of domains in cognitive science. It relies on a version of functionalist materialism, which holds that mental representations are actualized and modified by the individual holding the propositional attitude, and it challenges eliminative materialism and connectionism. It implies a strongly rationalist model of cognition in which many of the fundamentals of cognition are innate.

27.1 Presentation

The hypothesis applies to thoughts that have propositional content, and is not meant to describe everything that goes on in the mind. It appeals to the representational theory of thought to explain what those tokens actually are and how they behave. There must be a mental representation that stands in some unique relationship with the subject of the representation and has specific content. Complex thoughts get their semantic content from the content of the basic thoughts and the relations that they hold to each other. Thoughts can only relate to each other in ways that do not violate the syntax of thought. The syntax by means of which these two sub-parts are combined can be expressed in first-order predicate calculus.

The thought "John is tall" is clearly composed of two sub-parts, the concept of John and the concept of tallness, combined in a manner that may be expressed in first-order predicate calculus as a predicate 'T' ("is tall") that holds of the entity 'j' (John). A fully articulated proposal for what a LOT would have to take into account greater complexities such as quantification and propositional attitudes (the various attitudes people can have towards statements; for example I might *believe* or *see* or merely *suspect* that John is tall).

27.1.1 Precepts

1. There can be no higher cognitive processes without mental representation. The only plausible psychological models represent higher cognitive processes as representational and computational thought needs a representational system as an object upon which to compute. We must therefore attribute a representational system to organisms for cognition and thought to occur.

2. There is causal relationship between our intentions and our actions. Because mental states are structured in a way that causes our intentions to manifest themselves by what we do, there is a connection between how we view the world and ourselves and what we do.

27.2 Reception

Some philosophers have argued that our public language is our mental language, that a person who speaks English *thinks* in English. Others contend that people who do not know a public language (e.g. babies, aphasics) *can* think, and that therefore some form of mentalese must be present innately.

The notion that mental states are causally efficacious diverges from behaviorists like Gilbert Ryle, who held that there is no break between cause of mental state and effect of behavior. Rather, Ryle proposed that people act in some way because they are in a disposition to act in that way, that these causal mental states are representational. An objection to this point comes from John Searle in the form of biological naturalism, a nonrepresentational theory of mind that accepts the causal efficacy of mental states. Searle divides intentional states into low-level brain activity and high-level mental activity. The lower-level, nonrepresentational neurophysiological processes have causal power in intention and behavior rather than some higher-level mental representation.

Tim Crane, in his book *The Mechanical Mind*,[2] states that, while he agrees with Fodor, his reason is very different. A logical objection challenges LOTH's explanation of how sentences in natural languages get their meaning. That is the view that "Snow is white" is TRUE if and only if P is TRUE in the LOT, where P means the same thing in LOT as "Snow is white" means in the natural language. Any symbol manipulation is in need of some way of deriving what those symbols mean.[2] If the meaning of sentences is explained in terms of sentences in the LOT, then the meaning of sentences in LOT must get their meaning from somewhere else. There seems to be an infinite regress of sentences getting their meaning. Sentences in natural languages get their meaning from their users (speakers, writers).[2] Therefore sentences in mentalese must get their meaning from the way in which they are used by thinkers and so on *ad infinitum*. This regress is often called the homunculus regress.[2]

Daniel Dennett accepts that homunculi may be explained by other homunculi and denies that this would yield an infinite regress of homunculi. Each explanatory homunculus is "stupider" or more basic than the homunculus it explains but this regress is not infinite but bottoms out at a basic level that is so simple that it does not need interpretation.[2] John Searle points out that it still follows that the bottom-level homunculi are manipulating some sorts of symbols.

LOTH implies that the mind has some tacit knowledge of the logical rules of inference and the linguistic rules of syntax (sentence structure) and semantics (concept or word meaning).[2] If LOTH cannot show that the mind knows that it is following the particular set of rules in question, then the mind is not computational because it is not governed by computational rules.[2][3] Also, the apparent incompleteness of this set of rules in explaining behavior is pointed out. Many conscious beings behave in ways that are contrary to the rules of logic. Yet this irrational behavior is not accounted for by any rules, showing that there is at least some behavior that does not act in accordance with this set of rules.[2]

Another objection within representational theory of mind has to do with the relationship between propositional attitudes and representation. Dennett points out that a chess program can have the attitude of "wanting to get its queen out early," without having a representation or rule that explicitly states this. A multiplication program on a computer computes in the computer language of 1's and 0's, yielding representations that do not correspond with any propositional attitude.[3]

Susan Schneider has recently developed a version of LOT that departs from Fodor's approach in numerous ways. Schneider argues that Fodor's pessimism about the success of cognitive science is misguided, and she outlines an approach to LOT that integrates LOT with neuroscience. She also stresses that LOT that is not wedded to the extreme view that all concepts are innate. She fashions a new theory of mental symbols, and a related two-tiered theory of concepts, in which a concept's nature is determined by their LOT symbol type and their meaning.[4]

27.3   Relation to connectionism

Connectionism is a recent applied approach to artificial intelligence that often accepts a lot of the same theoretical framework that LOTH accepts, namely that mental states are computational and causally efficacious and very often that they are representational. However, connectionism stresses the possibility of thinking machines, most often realized as neural networks, an inter-connectional set of nodes, and describes mental states as able to create memory by modifying the strength of these connections over time. Some popular types of neural networks are interpretations of units, and learning algorithm. "Units" can be interpreted as neurons or groups of neurons. A learning algorithm is such that, over time, a change in connection weight is possible, allowing networks to modify their connections. Connectionist neural networks are able to change over time via their activation. An activation is a numerical value that represents any aspect of a unit that a neural network has at any time. Activation spreading is the spreading or taking over of other over time of the activation to all other units connected to the activated unit.

Since connectionist models can change over time, supporters of connectionism claim that it can solve the problems that LOTH brings to classical AI. These problems are those that show that machines with a LOT syntactical framework very often are much better at solving problems and storing data than human minds, yet much worse at things that the human mind is quite adept at such as recognizing facial expressions and objects in photographs and understanding nuanced gestures.[2] Fodor defends LOTH by arguing that a connectionist model is just some realization or implementation of the classical computational theory of mind and therein necessarily employs a symbol-manipulating LOT.

Fodor and Zenon Pylyshyn use the notion of cognitive architecture in their defense. Cognitive architecture is the set of basic functions of an organism with representational input and output. They argue that it is a law of nature that cognitive capacities are productive, systematic and inferentially coherent - they have the ability to produce and understand sentences of a certain structure if they can understand one sentence of that structure.[5] A cognitive model must have a cognitive architecture that explains these laws and properties in some way that is compatible with the scientific method. Fodor and Pylyshyn say that cognitive architecture can only explain the property of systematicity by appealing to a system of representations and that connectionism either employs a cognitive architecture of representations or else does not. If it does, then connectionism uses LOT. If it does not then it is empirically false.[3]

Connectionists have responded to Fodor and Pylyshyn by denying that connectionism uses LOT, by denying that cognition is essentially a function that uses representational input and output or denying that systematicity is a law of nature that rests on representation.

27.4   Empirical testing

Since LOTH came to be it has been empirically tested. Not all experiments have confirmed the hypothesis;

- In 1971, Roger Shepard and Jacqueline Metzler tested Pylyshyn's particular hypothesis that all symbols are understood by the mind in virtue of their fundamental mathematical descriptions.[6] Shepard and Metzler's experiment consisted of showing a group of subjects a 2-D line drawing of a 3-D object, and then that same object at some rotation. According to Shepard and Metzler, if Pylyshyn were correct, then the amount of time it took to identify the object as the same object would not depend on the degree of rotation of the object. Their finding that the time taken to recognize the object was proportional to its rotation contradicts this hypothesis.

- There may be a connection between prior knowledge of what relations hold between objects in the world and the time it takes subjects to recognize the same objects. For example, it is more likely that subjects will not recognize a hand that is rotated in such a way that it would be physically impossible for an actual hand. It has since also been empirically tested and supported that the mind might better manipulate mathematical descriptions in topographical wholes. These findings have illuminated what the mind is not doing in terms of how it manipulates symbols.

27.5   See also

- Private language argument

- Universal grammar
- Psycholinguistics
- Psychological nativism

27.6   References

[1] Stanford Encyclopedia of Philosophy http://plato.stanford.edu/entries/language-thought/

[2] Crane, Tim (2005). *The mechanical mind : a philosophical introduction to minds, machines and mental representation* (2nd, repr. ed.). London: Routledge. ISBN 978-0-415-29031-9.

[3] Murat Aydede (2004-07-27). "The Language of Thought Hypothesis".

[4] Schneider, Susan (2011). *The Language of Thought: a New Direction*. Boston: Mass: MIT Press.

[5] James Garson (2010-07-27). "Connectionism".

[6] Shepard, Roger N.; Metzler, Jacqueline (1971-02-19). "Mental Rotation of Three-Dimensional Objects". *Science* **171** (3972): 701–703. doi:10.1126/science.171.3972.701. PMID 5540314.

- Ravenscroft, Ian, *Philosophy of mind*. Oxford University press, 2005. pp 91.
- Fodor, Jerry A., *The Language Of Thought*. Crowell Press, 1975. pp 214.
- John R. Searle (June 29, 1972). "Chomsky's Revolution in Linguistics". *New York Review of Books*.

27.7   External links

- The Language of Thought Hypothesis at The Internet Encyclopedia of Philosophy.
- The Language of Thought Hypothesis at The Stanford Encyclopedia of Philosophy.
- Language of Thought - By Larry Kaye.
- Revealing The Language Of Thought - By Brent Silby
- Jerry Fodor Homepage
- The Language Of Thought Hypothesis: State Of The Art - By Murat Aydede

Chapter 28

Automated theorem proving

Argonne National Laboratory was a leader in automated theorem proving from the 1960s to the 2000s

Automated theorem proving (also known as **ATP** or **automated deduction**) is a subfield of automated reasoning and mathematical logic dealing with proving mathematical theorems by computer programs. Automated reasoning over mathematical proof was a major impetus for the development of computer science.

28.1   Logical foundations

While the roots of formalised logic go back to Aristotle, the end of the 19th and early 20th centuries saw the development of modern logic and formalised mathematics. Frege's *Begriffsschrift* (1879) introduced both a complete propositional calculus and what is essentially modern predicate logic.[1] His *Foundations of Arithmetic*, published 1884,[2] expressed (parts of) mathematics in formal logic. This approach was continued by Russell and Whitehead in their influential *Principia Mathematica*, first published 1910–1913,[3] and with a revised second edition in 1927.[4] Russell and Whitehead thought they could derive all mathematical truth using axioms and inference rules of formal logic, in principle opening up the process to automatisation. In 1920, Thoralf Skolem simplified a previous result by Leopold Löwenheim, leading to the Löwenheim–Skolem theorem and, in 1930, to the notion of a Herbrand universe and a Herbrand interpretation that allowed (un)satisfiability of first-order formulas (and hence the validity of a theorem) to be reduced to (potentially infinitely many) propositional satisfiability problems.[5]

In 1929, Mojżesz Presburger showed that the theory of natural numbers with addition and equality (now called Presburger arithmetic in his honor) is decidable and gave an algorithm that could determine if a given sentence in the language was true or false.[6][7] However, shortly after this positive result, Kurt Gödel published *On Formally Undecidable Propositions of Principia Mathematica and Related Systems* (1931), showing that in any sufficiently strong axiomatic system there are true statements which cannot be proved in the system. This topic was further developed in the 1930s by Alonzo Church and Alan Turing, who on the one hand gave two independent but equivalent definitions of computability, and on the other gave concrete examples for undecidable questions.

28.2   First implementations

Shortly after World War II, the first general purpose computers became available. In 1954, Martin Davis programmed Presburger's algorithm for a JOHNNIAC vacuum tube computer at the Princeton Institute for Advanced Study. According to Davis, "Its great triumph was to prove that the sum of two even numbers is even".[7][8] More ambitious was the Logic Theory Machine, a deduction system for the propositional logic of the *Principia Mathematica*, developed by Allen Newell, Herbert A. Simon and J. C. Shaw. Also running on a JOHNNIAC, the Logic Theory Machine constructed proofs from a small set of propositional axioms and three deduction rules: modus ponens, (propositional) variable substitution, and the replacement of formulas by their definition. The system used heuristic guidance, and managed to prove 38 of the first 52 theorems of the *Principia*.[7]

The "heuristic" approach of the Logic Theory Machine tried to emulate human mathematicians, and could not guarantee that a proof could be found for every valid theorem even in principle. In contrast, other, more systematic algorithms achieved, at least theoretically, completeness for first-order logic. Initial approaches relied on the results of Herbrand and Skolem to convert a first-order formula into successively larger sets of propositional formulae by instantiating variables with terms from the Herbrand universe. The propositional formulas could then be checked for unsatisfiability using a number of methods. Gilmore's program used conversion to disjunctive normal form, a form in which the satisfiability of a formula is obvious.[7][9]

28.3   Decidability of the problem

Depending on the underlying logic, the problem of deciding the validity of a formula varies from trivial to impossible. For the frequent case of propositional logic, the problem is decidable but Co-NP-complete, and hence only exponential-time algorithms are believed to exist for general proof tasks. For a first order predicate calculus, Gödel's completeness theorem states that the theorems (provable statements) are exactly the logically valid well-formed formulas, so identifying valid formulas is recursively enumerable: given unbounded resources, any valid formula can eventually be proven. However, *invalid* formulas (those that are *not* entailed by a given theory), cannot always be recognized.

The above applies to first order theories, such as Peano Arithmetic. However, for a specific model that may be described by a first order theory, some statements may be true but undecidable in the theory used to describe the model. For example, by Gödel's incompleteness theorem, we know that any theory whose proper axioms are true for the natural numbers cannot prove all first order statements true for the natural numbers, even if the list of proper axioms is allowed to be infinite enumerable. It follows that an automated theorem prover will fail to terminate while searching for a proof precisely when the statement being investigated is undecidable in the theory being used, even if it is true in the model of interest. Despite this theoretical limit, in practice, theorem provers can solve many hard problems, even in models that are not fully described by any first order theory (such as the integers).

28.4 Related problems

A simpler, but related, problem is **proof verification**, where an existing proof for a theorem is certified valid. For this, it is generally required that each individual proof step can be verified by a primitive recursive function or program, and hence the problem is always decidable.

Since the proofs generated by automated theorem provers are typically very large, the problem of proof compression is crucial and various techniques aiming at making the prover's output smaller, and consequently more easily understandable and checkable, have been developed.

Proof assistants require a human user to give hints to the system. Depending on the degree of automation, the prover can essentially be reduced to a proof checker, with the user providing the proof in a formal way, or significant proof tasks can be performed automatically. Interactive provers are used for a variety of tasks, but even fully automatic systems have proved a number of interesting and hard theorems, including at least one that has eluded human mathematicians for a long time, namely the Robbins conjecture.[10][11] However, these successes are sporadic, and work on hard problems usually requires a proficient user.

Another distinction is sometimes drawn between theorem proving and other techniques, where a process is considered to be theorem proving if it consists of a traditional proof, starting with axioms and producing new inference steps using rules of inference. Other techniques would include model checking, which, in the simplest case, involves brute-force enumeration of many possible states (although the actual implementation of model checkers requires much cleverness, and does not simply reduce to brute force).

There are hybrid theorem proving systems which use model checking as an inference rule. There are also programs which were written to prove a particular theorem, with a (usually informal) proof that if the program finishes with a certain result, then the theorem is true. A good example of this was the machine-aided proof of the four color theorem, which was very controversial as the first claimed mathematical proof which was essentially impossible to verify by humans due to the enormous size of the program's calculation (such proofs are called non-surveyable proofs). Another example would be the proof that the game Connect Four is a win for the first player.

28.5 Industrial uses

Commercial use of automated theorem proving is mostly concentrated in integrated circuit design and verification. Since the Pentium FDIV bug, the complicated floating point units of modern microprocessors have been designed with extra scrutiny. AMD, Intel and others use automated theorem proving to verify that division and other operations are correctly implemented in their processors.

28.6 First-order theorem proving

In the late 1960s agencies funding research in automated deduction began to emphasize the need for practical applications. One of the first fruitful areas was that of program verification whereby first-order theorem provers were applied to the problem of verifying the correctness of computer programs in languages such as Pascal, Ada, Java etc. Notable among early program verification systems was the Stanford Pascal Verifier developed by David Luckham at Stanford University. This was based on the Stanford Resolution Prover also developed at Stanford using J.A. Robinson's resolution Principle. This was the first automated deduction system to demonstrate an ability to solve mathematical problems that were announced in the Notices of the American Mathematical Society before solutions were formally published.

First-order theorem proving is one of the most mature subfields of automated theorem proving. The logic is expressive enough to allow the specification of arbitrary problems, often in a reasonably natural and intuitive way. On the other hand, it is still semi-decidable, and a number of sound and complete calculi have been developed, enabling *fully* automated systems. More expressive logics, such as higher order logics, allow the convenient expression of a wider range of problems than first order logic, but theorem proving for these logics is less well developed.

28.7   Benchmarks and competitions

The quality of implemented systems has benefited from the existence of a large library of standard benchmark examples — the Thousands of Problems for Theorem Provers (TPTP) Problem Library[12] — as well as from the CADE ATP System Competition (CASC), a yearly competition of first-order systems for many important classes of first-order problems.

Some important systems (all have won at least one CASC competition division) are listed below.

- E is a high-performance prover for full first-order logic, but built on a purely equational calculus, developed primarily in the automated reasoning group of Technical University of Munich.

- Otter, developed at the Argonne National Laboratory, is based on first-order resolution and paramodulation. Otter has since been replaced by Prover9, which is paired with Mace4.

- SETHEO is a high-performance system based on the goal-directed model elimination calculus. It is developed in the automated reasoning group of Technical University of Munich. E and SETHEO have been combined (with other systems) in the composite theorem prover E-SETHEO.

- Vampire is developed and implemented at Manchester University by Andrei Voronkov and Krystof Hoder, formerly also by Alexandre Riazanov. It has won the CADE ATP System Competition in the most prestigious CNF (MIX) division for eleven years (1999, 2001–2010).

- Waldmeister is a specialized system for unit-equational first-order logic. It has won the CASC UEQ division for the last fourteen years (1997–2010).

- SPASS is a first order logic theorem prover with equality. This is developed by the research group Automation of Logic, Max Planck Institute for Computer Science.

28.8   Popular techniques

- First-order resolution with unification

- Lean theorem proving

- Model elimination

- Method of analytic tableaux

- Superposition and term rewriting

- Model checking

- Mathematical induction

- Binary decision diagrams

- DPLL

- Higher-order unification

28.9 Comparison

See also: Proof assistant#Comparison and Category:Theorem proving software systems

28.9.1 Free software

- Alt-Ergo
- Automath
- CVC
- E
- Gödel-machines
- iProver
- IsaPlanner
- KED theorem prover
- LCF
- LoTREC
- MetaPRL
- NuPRL
- Paradox
- Simplify (GPL'ed since 5/2011)
- Twelf
- SPARK (programming language)

28.9.2 Proprietary software

- Acumen RuleManager (commercial product)
- ALLIGATOR
- CARINE
- KIV
- Prover Plug-In (commercial proof engine product)
- ProverBox
- ResearchCyc
- Spear modular arithmetic theorem prover

28.10   Notable people

- Leo Bachmair, co-developer of the superposition calculus.

- Woody Bledsoe, artificial intelligence pioneer.

- Robert S. Boyer, co-author of the Boyer-Moore theorem prover, co-recipient of the Herbrand Award 1999.

- Alan Bundy, University of Edinburgh, meta-level reasoning for guiding inductive proof, proof planning and recipient of 2007 IJCAI Award for Research Excellence, Herbrand Award, and 2003 Donald E. Walker Distinguished Service Award.

- William McCune Argonne National Laboratory, author of Otter, the first high-performance theorem prover. Many important papers, recipient of the Herbrand Award 2000.

- Hubert Comon, CNRS and now ENS Cachan. Many important papers.

- Robert Lee Constable, Cornell University. Important contributions to type theory, NuPRL.

- Martin Davis, author of the "Handbook of Artificial Reasoning", co-inventor of the DPLL algorithm, recipient of the Herbrand Award 2005.

- Branden Fitelson University of California at Berkeley. Work in automated discovery of shortest axiomatic bases for logic systems.

- Harald Ganzinger, co-developer of the superposition calculus, head of the MPI Saarbrücken, recipient of the Herbrand Award 2004 (posthumous).

- Michael Genesereth, Stanford University professor of Computer Science.

- Keith Goolsbey chief developer of the Cyc inference engine.

- Michael J. C. Gordon led the development of the HOL theorem prover.

- Gérard Huet Term rewriting, HOL logics, Herbrand Award 1998.

- Robert Kowalski developed the connection graph theorem-prover and SLD resolution, the inference engine that executes logic programs.

- Donald W. Loveland Duke University. Author, co-developer of the DPLL-procedure, developer of model elimination, recipient of the Herbrand Award 2001.

- David Luckham Stanford University, Developed the Stanford Resolution Theorem Prover 1968, the first automated deduction system used to solve problems announced in the Notices of the AMS, and subsequently developed the Stanford Pascal Verifier, the first program verification system for Pascal, and a widely distributed program verification system, 1968-75

- Norman Megill, developer of Metamath, and maintainer of its site at metamath.org, an online database of automatically verified proofs.

- J Strother Moore, co-author of the Boyer–Moore theorem prover, co-recipient of the Herbrand Award 1999.

- Robert Nieuwenhuis University of Barcelona. Co-developer of the superposition calculus.

- Tobias Nipkow of the Technical University of Munich, contributions to (higher-order) rewriting, co-developer of the Isabelle proof assistant

- Ross Overbeek Argonne National Laboratory. Founder of The Fellowship for Interpretation of Genomes

- Lawrence C. Paulson of the University of Cambridge, work on higher-order logic system, co-developer of the Isabelle Theorem Prover

- David Plaisted University of North Carolina at Chapel Hill. Complexity results, contributions to rewriting and completion, instance-based theorem proving.

- John Rushby Program Director – SRI International[13]

- J. Alan Robinson Syracuse University. Developed original resolution and unification based first order theorem proving, co-editor of the "Handbook of Automated Reasoning", recipient of the Herbrand Award 1996

- Jürgen Schmidhuber Work on Gödel Machines: Self-Referential Universal Problem Solvers Making Provably Optimal Self-Improvements

- Stephan Schulz, E theorem Prover.

- Natarajan Shankar SRI International, work on decision procedures, *little engines of proof*, co-developer of PVS.

- Mark Stickel SRI International. Recipient of the Herbrand Award 2002.

- Geoff Sutcliffe University of Miami. Maintainer of the TPTP collection, an organizer of the CADE annual contest.

- Dolph Ulrich Purdue, Work on automated discovery of shortest axiomatic bases for systems.

- Robert Veroff University of New Mexico. Many important papers.

- Andrei Voronkov Developer of Vampire and Co-Editor of the "Handbook of Automated Reasoning"

- Larry Wos Argonne National Laboratory. (Otter) Many important papers. Very first Herbrand Award winner (1992)

- Wen-Tsun Wu Work in geometric theorem proving: Wu's method, Herbrand Award 1997

- Christoph Weidenbach, author of SPASS, automated theorem prover.

28.11   See also

- Symbolic computation

- Computer-aided proof

- Automated reasoning

- Formal verification

- Logic programming

- Proof checking

- Model checking

- Proof complexity

- Computer algebra system

- Program analysis (computer science)

- General Problem Solver

- Metamath language for formalized mathematics

28.12 Notes

[1] Frege, Gottlob (1879). *Begriffsschrift*. Verlag Louis Neuert.

[2] Frege, Gottlob (1884). *Die Grundlagen der Arithmetik* (PDF). Breslau: Wilhelm Kobner.

[3] Bertrand Russell; Alfred North Whitehead (1910–1913). *Principia Mathematica* (1st ed.). Cambridge University Press.

[4] Bertrand Russell; Alfred North Whitehead (1927). *Principia Mathematica* (2nd ed.). Cambridge University Press.

[5] Herbrand, Jaques (1930). *Recherches sur la théorie de la démonstration*.

[6] Presburger, Mojżesz (1929). "Über die Vollständigkeit eines gewissen Systems der Arithmetik ganzer Zahlen, in welchem die Addition als einzige Operation hervortritt". *Comptes Rendus du I congrès de Mathématiciens des Pays Slaves* (Warszawa): 92–101.

[7] Davis, Martin (2001), "The Early History of Automated Deduction", in Robinson, Alan; Voronkov, Andrei, *Handbook of Automated Reasoning* **1**, Elsevier)

[8] Bibel, Wolfgang (2007). "Early History and Perspectives of Automated Deduction" (PDF). *KI 2007*. LNAI (Springer) (4667): 2–18. Retrieved 2 September 2012.

[9] Gilmore, Paul (1960). "A proof procedure for quantification theory: its justification and realisation". *IBM Journal of Research and Development* **4**: 28–35. doi:10.1147/rd.41.0028.

[10] W.W. McCune (1997). "Solution of the Robbins Problem". *Journal of Automated Reasoning* **19** (3).

[11] Gina Kolata (December 10, 1996). "Computer Math Proof Shows Reasoning Power". The New York Times. Retrieved 2008-10-11.

[12] Sutcliffe, Geoff. "The TPTP Problem Library for Automated Theorem Proving". Retrieved 8 September 2012.

[13] "SRI International Computer Science Laboratory – John Rushby". SRI International. Retrieved 22 September 2012.

28.13 References

- Chin-Liang Chang; Richard Char-Tung Lee (1973). *Symbolic Logic and Mechanical Theorem Proving*. Academic Press.

- Loveland, Donald W. (1978). *Automated Theorem Proving: A Logical Basis. Fundamental Studies in Computer Science Volume 6*. North-Holland Publishing.

- Luckham, David (1990). *Programming with Specifications: An Introduction to Anna, A Language for Specifying Ada Programs*. Springer-Verlag Texts and Monographs in Computer Science, 421 pp.

- Gallier, Jean H. (1986). *Logic for Computer Science: Foundations of Automatic Theorem Proving*. Harper & Row Publishers (Available for free download).

- Duffy, David A. (1991). *Principles of Automated Theorem Proving*. John Wiley & Sons.

- Wos, Larry; Overbeek, Ross; Lusk, Ewing; Boyle, Jim (1992). *Automated Reasoning: Introduction and Applications* (2nd ed.). McGraw–Hill.

- Alan Robinson and Andrei Voronkov (eds.), ed. (2001). *Handbook of Automated Reasoning Volume I & II*. Elsevier and MIT Press.

- Fitting, Melvin (1996). *First-Order Logic and Automated Theorem Proving* (2nd ed.). Springer.

Chapter 29

Mathematical fallacy

In mathematics, certain kinds of mistaken proof are often exhibited, and sometimes collected, as illustrations of a concept of **mathematical fallacy**. There is a distinction between a simple *mistake* and a *mathematical fallacy* in a proof: a mistake in a proof leads to an **invalid proof** just in the same way, but in the best-known examples of mathematical fallacies, there is some concealment in the presentation of the proof. For example, the reason validity fails may be a division by zero that is hidden by algebraic notation. There is a striking quality of the mathematical fallacy: as typically presented, it leads not only to an absurd result, but does so in a crafty or clever way.[1] Therefore these fallacies, for pedagogic reasons, usually take the form of spurious proofs of obvious contradictions. Although the proofs are flawed, the errors, usually by design, are comparatively subtle, or designed to show that certain steps are conditional, and should not be applied in the cases that are the exceptions to the rules.

The traditional way of presenting a mathematical fallacy is to give an invalid step of deduction mixed in with valid steps, so that the meaning of fallacy is here slightly different from the logical fallacy. The latter applies normally to a form of argument that is not a genuine rule of logic, where the problematic mathematical step is typically a correct rule applied with a tacit wrong assumption. Beyond pedagogy, the resolution of a fallacy can lead to deeper insights into a subject (such as the introduction of Pasch's axiom of Euclidean geometry[2] and the five color theorem of graph theory). *Pseudaria*, an ancient lost book of false proofs, is attributed to Euclid.[3]

Mathematical fallacies exist in many branches of mathematics. In elementary algebra, typical examples may involve a step where division by zero is performed, where a root is incorrectly extracted or, more generally, where different values of a multiple valued function are equated. Well-known fallacies also exist in elementary Euclidean geometry and calculus.

29.1 Howlers

Examples exist of *mathematically **correct** results derived by **incorrect** lines of reasoning*. Such an argument, however true the conclusion, is mathematically invalid and is commonly known as a **howler**. Consider for instance the calculation (anomalous cancellation):

$$\frac{16}{64} = \frac{1\not6}{\not64} = \frac{1}{4}.$$

Although the conclusion $\frac{16}{64} = \frac{1}{4}$ is correct, there is a fallacious, invalid cancellation in the middle step. Bogus proofs, calculations, or derivations constructed to produce a correct result in spite of incorrect logic or operations were termed *howlers* by Maxwell.[4] Outside the field of mathematics the term "*howler*" has various meanings, generally less specific.

29.2   Division by zero

The division-by-zero fallacy has many variants. The following example uses division by zero to "prove" that $2 = 1$, but can be modified to prove that any number equals any other number.

1. Let a and b be equal non-zero quantities

   $a = b$

2. Multiply by a

   $a^2 = ab$

3. Subtract b^2

   $a^2 - b^2 = ab - b^2$

4. Factor both sides; the left factors as a difference of squares, the right is factored through its greatest common divisor)

   $(a - b)(a + b) = b(a - b)$

5. Divide out $(a - b)$

   $a + b = b$

6. Observing that $a = b$

   $b + b = b$

7. Combine like terms on the left

   $2b = b$

8. Divide by the non-zero b

   $2 = 1$

Q.E.D.[5]

The fallacy is in line 5: the progression from line 4 to line 5 involves division by $a - b$, which is zero since a equals b. Since division by zero is undefined, the argument is invalid.

29.3   Multivalued functions

Many functions do not have a unique inverse. For instance squaring a number gives a unique value, but there are two possible square roots of a positive number. The square root is multivalued. One value can be chosen by convention as the principal value, in the case of the square root the non-negative value is the principal value, but there is no guarantee that the square root function given by this principal value of the square of a number will be equal to the original number, e.g. the square root of the square of -2 is 2.

29.4 Calculus

Calculus as the mathematical study of infinitesimal change and limits can lead to mathematical fallacies if the properties of integrals and differentials are ignored. For instance, a naive use of integration by parts can be used to give a false proof that $0 = 1$.[6] Letting $u = \frac{1}{\log x}$ and $dv = \frac{dx}{x}$, we may write:

$$\int \frac{1}{x \log x}\, dx = 1 + \int \frac{1}{x \log x}\, dx$$

after which the antiderivatives may be cancelled yielding $0 = 1$. The problem is that antiderivatives are only defined up to a constant and shifting them by 1 or indeed any number is allowed. The error really comes to light when we introduce arbitrary integration limits a and b.

$$\int_a^b \frac{1}{x \log x}\, dx = 1|_a^b + \int_a^b \frac{1}{x \log x}\, dx = 0 + \int_a^b \frac{1}{x \log x}\, dx = \int_a^b \frac{1}{x \log x}\, dx$$

Since the difference between two values of a constant function vanishes, the same definite integral appears on both sides of the equation.

29.5 Power and root

Fallacies involving disregarding the rules of elementary arithmetic through an incorrect manipulation of the radical. For complex numbers the failure of power and logarithm identities has led to many fallacies.

29.5.1 Positive and negative roots

Proof of

$$5 = 4$$

1. Start from

$$-20 = -20$$

2. Write this as

$$25 - 45 = 16 - 36$$

3. Rewrite as

$$5^2 - 5 * 9 = 4^2 - 4 * 9$$

4. Add $81/4$ on both sides:

$$5^2 - 5 * 9 + 81/4 = 4^2 - 4 * 9 + 81/4$$

5. These are perfect squares:

$$(5 - 9/2)^2 = (4 - 9/2)^2$$

6. Take the square root of both sides:

$$5 - 9/2 = 4 - 9/2$$

7. Add $9/2$ on both sides:

$$5 = 4$$

Q.E.D.[7]

The fallacy is in line 6: $a^2=b^2$ only implies $a=b$ if a and b have the same sign, which not the case here. In this case it implies $a=-b$ and should read

$$5 - 9/2 = -(4 - 9/2).$$

29.5.2   Square roots of negative numbers

Invalid proofs utilizing powers and roots are often of the following kind:[8]

$$1 = \sqrt{1} = \sqrt{(-1)(-1)} = \sqrt{-1}\sqrt{-1} = i \cdot i = -1.$$

The fallacy is that the rule $\sqrt{xy} = \sqrt{x}\sqrt{y}$ is generally valid only if both x and y are positive (when dealing with real numbers), which is not the case here.

Although the fallacy is easily detected here, sometimes it is concealed more effectively in notation. For instance,[9] consider the equation

$$\cos^2 x = 1 - \sin^2 x$$

which holds as a consequence of the Pythagorean theorem. Then, by taking a square root,

$$\cos x = (1 - \sin^2 x)^{\frac{1}{2}}$$

so that

$$1 + \cos x = 1 + (1 - \sin^2 x)^{\frac{1}{2}}.$$

But evaluating this when $x = \pi$ implies

$$1 - 1 = 1 + (1 - 0)^{\frac{1}{2}}$$

or

$$0 = 2$$

which is incorrect.

The error in each of these examples fundamentally lies in the fact that any equation of the form

$$x^2 = a^2$$

has two solutions, provided $a \neq 0$,

$$x = \pm a$$

and it is essential to check which of these solutions is relevant to the problem at hand.[10] In the above fallacy, the square root that allowed the second equation to be deduced from the first is valid only when cos x is positive. In particular, when x is set to π, the second equation is rendered invalid.

Another example of this kind of fallacy, where the error is immediately detectable, is the following invalid proof that -2 $= 2$. Letting $x = -2$, and then squaring gives

$$x^2 = 4$$

whereupon taking a square root implies

$$x = \sqrt{4} = 2,$$

so that $x = -2 = 2$, which is absurd. Clearly when the square root was extracted, it was the *negative* root -2, rather than the *positive* root, that was relevant for the particular solution in the problem.

Alternatively, imaginary roots are obfuscated in the following:

$$\sqrt{-1} = (-1)^{\frac{2}{4}} = ((-1)^2)^{\frac{1}{4}} = 1^{\frac{1}{4}} = 1$$

The error here lies in the last equality, where we are ignoring the other fourth roots of 1,[11] which are -1, i and $-i$ (where i is the imaginary unit). Seeing as we have squared our figure and then taken roots, we cannot always assume that all the roots will be correct. So the correct fourth roots are i and $-i$, which are the imaginary numbers defined to square to -1.

29.5.3   Complex exponents

When a number is raised to a complex power, the result is not uniquely defined (see Failure of power and logarithm identities). If this property is not recognized, then errors such as the following can result:

$$e^{2\pi i} = 1$$
$$(e^{2\pi i})^i = 1^i$$
$$e^{-2\pi} = 1$$

The error here is that the rule of multiplying exponents as when going to the third line does not apply unmodified with complex exponents, even if when putting both sides to the power i only the principal value is chosen. When treated as multivalued functions, both sides produce the same set of values, being $\{e^{2\pi n} \mid n \in \mathbb{Z}\}$.

29.6   Geometry

Many mathematical fallacies in geometry arise from using in an additive equality involving oriented quantities (such adding vectors along a given line or adding oriented angles in the plane) a valid identity, but which fixes only the absolute value of (one of) these quantities. This quantity is then incorporated into the equation with the wrong orientation, so as to produce an absurd conclusion. This wrong orientation is usually suggested implicitly by supplying an imprecise diagram of the situation, where relative positions of points or lines are chosen in a way that is actually impossible under the hypotheses of the argument, but non-obviously so. Such a fallacy is easy to expose by drawing a precise picture of the situation, in which some relative positions will be different form those in the provided diagram. In order to avoid such fallacies, a correct geometric argument using addition or subtraction of distances or angles should always prove that quantities are being incorporated with their correct orientation.

29.6.1   Fallacy of the isosceles triangle

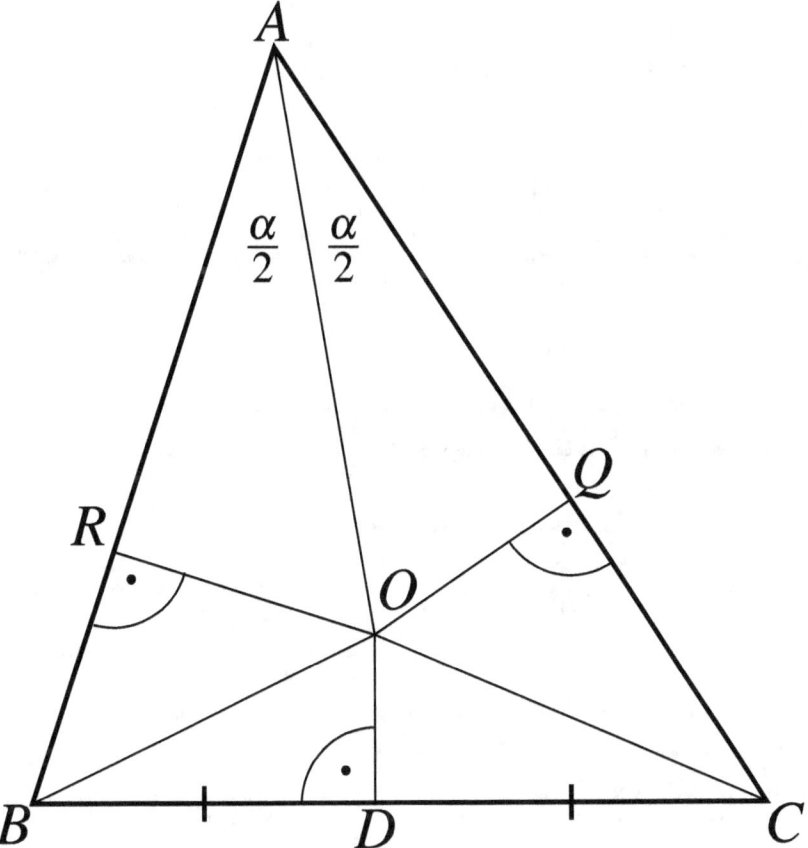

The fallacy of the isosceles triangle, from (Maxwell 1959, Chapter II, § 1), purports to show that every triangle is isosceles, meaning that two sides of the triangle are congruent. This fallacy has been attributed to Lewis Carroll.[12]

Given a triangle $\triangle ABC$, prove that AB = AC:

1. Draw a line bisecting ∠A

2. Draw the perpendicular bisector of segment BC, which bisects BC at a point D

3. Let these two lines meet at a point O.

4. Draw line OR perpendicular to AB, line OQ perpendicular to AC

5. Draw lines OB and OC

6. By AAS, $\triangle RAO \cong \triangle QAO$ (∠ORA = ∠OQA = 90; ∠RAO = ∠QAO; AO=AO (COMMON SIDE))

7. By RHS,[13] $\triangle ROB \cong \triangle QOC$

8. Thus, AR = AQ, RB = QC, and AB = AR + RB = AQ + QC = AC

Q.E.D.

As a corollary, one can show that all triangles are equilateral, by showing that AB = BC and AC = BC in the same way.

The error in the proof is the assumption in the diagram that the point O is *inside* the triangle. In fact, O always lies at the circumcircle of the $\triangle ABC$ (except for isosceles and equilateral triangles where AO and OD coincides . Furthermore, it can be shown that, if AB is longer than AC, then R will lie *within* AB, while Q will lie *outside* of AC (and vice versa). (Any diagram drawn with sufficiently accurate instruments will verify the above two facts.) Because of this, AB is still AR + RB, but AC is actually AQ – QC; and thus the lengths are not necessarily the same.

29.7 Proof by induction

There exist several fallacious proofs by induction in which one of the components, basis case or inductive step, is incorrect. Intuituvely, proofs by induction work by arguing that, if a statement is true in one case, it is true in the next case, and hence by repeatedly applying this it can be shown to be true for all cases. This "proof" shows that all horses are the same colour.

1. Let us say that any group of N horses is all of the same colour.

2. If we remove a horse from the group, we have a group of N - 1 horses of the same colour. If we add another horse, we have another group of N horses. By our previous assumption, all the horses are of the same colour in this new group, since it is a group of N horses.

3. Thus we have constructed two groups of N horses all of the same colour, with N - 1 horses in common. Since these two groups have some horses in common, the two groups must be of the same colour as each other.

4. Therefore combining all the horses used, we have a group of N + 1 horses of the same colour.

5. Thus if any N horses are all the same colour, any N + 1 horses are the same colour.

6. This is clearly true for N = 1 (i.e. one horse is a group where all the horses are the same colour). Thus, by induction, N horses are the same colour for any positive integer N. i.e. all horses are the same colour.

The fallacy in this proof arises in line 3. For N = 1, the two groups of horses have N − 1 = 0 horses in common, and thus are not necessarily the same colour as each other, so the group of N + 1 = 2 horses is not necessarily all of the same colour. The implication "Every N horses are of the same color, then N+1 horses are of the same color" works for any N greater than one, but fails to be true when N=1. The basis case is correct, but the induction step has a fundamental flaw.

29.8   See also

- List of incomplete proofs
- Paradox
- Proof by intimidation

29.9   Notes

[1] Maxwell 1959, p. 9

[2] Maxwell 1959

[3] Heath & Helberg 1908, Chapter II, §I

[4] Maxwell 1959

[5] Heuser, Harro (1989), *Lehrbuch der Analysis - Teil 1* (6th ed. ed.), Teubner, p. 51, ISBN 978-3-8351-0131-9

[6] Barbeau, Ed (1990), "Fallacies, Flaws and Flimflam #19: Dolt's Theorem", *The College Mathematics Journal* **21** (3): 216–218

[7] Frohlichstein, Jack (1967). *Mathematical Fun, Games and Puzzles* (illustrated ed.). Courier Corporation. p. 207. ISBN 0-486-20789-7., Extract of page 207

[8] Maxwell 1959, Chapter VI, §I.2

[9] Maxwell 1959, Chapter VI, §I.1

[10] Maxwell 1959, Chapter VI, §II

[11] In general, the expression $\sqrt[n]{1}$ evaluates to n complex numbers, called the nth roots of unity.

[12] Robin Wilson (2008), *Lewis Carroll in Numberland*, Penguin Books, pp. 169–170, ISBN 978-0-14-101610-8

[13] Hypotenuse-leg congruence

29.10   References

- Barbeau, Edward J. (2000), *Mathematical fallacies, flaws, and flimflam*, MAA Spectrum, Mathematical Association of America, ISBN 978-0-88385-529-4, MR 1725831.

- Bunch, Bryan (1997), *Mathematical fallacies and paradoxes*, New York: Dover Publications, ISBN 978-0-486-29664-7, MR 1461270.

- Heath, Sir Thomas Little; Heiberg, Johan Ludvig (1908), *The thirteen books of Euclid's Elements, Volume 1*, The University Press.

- Maxwell, E. A. (1959), *Fallacies in mathematics*, Cambridge University Press, ISBN 0-521-05700-0, MR 0099907.

29.11   External links

- Invalid proofs at Cut-the-knot (including literature references)
- Classic fallacies with some discussion
- More invalid proofs from AhaJokes.com
- Math jokes including an invalid proof

Chapter 30

List of incomplete proofs

This page lists notable examples of incomplete published mathematical proofs. Most of these were accepted as correct for several years but later discovered to contain gaps. There are both examples where a complete proof was later found and where the alleged result turned out to be false.

Lecat (1935) is a list over a hundred pages long of errors made by mathematicians.

30.1 Examples

This section lists examples of proofs that were published and accepted as complete before a gap or error was found in them. It does not include any of the many incomplete attempted solutions by amateurs of famous problems such as Fermat's last theorem or the squaring of the circle. It also does not include unpublished preprints that were withdrawn because an error was found before publication.

The examples are arranged roughly in order of the publication date of the incomplete proof. Several of the examples on the list were taken from answers to questions on the MathOverflow site, listed in the external links below. The examples use the following symbols:

- Result is correct and was later rigorously proved.

- Result is wrong as stated, but a modified version was later rigorously proved.

- Status of the result is unclear

- Result is wrong

- Euclid's Elements. Euclid's proofs are essentially correct, but strictly speaking sometimes contain gaps because he tacitly uses some unstated assumptions, such as the existence of intersection points. In 1899 Hilbert gave a complete set of (second order) axioms for Euclidean geometry, called Hilbert's axioms, and between 1926 and 1959 Tarski gave some complete sets of first order axioms, called Tarski's axioms.

- Infinitesimals. In the 18th century there was widespread use of infinitesimals in calculus, though these were not really well defined. Calculus was put on firm foundations in the 19th century, and Robinson put infinitesimals in a rigorous basis with the introduction of nonstandard analysis in the 20th century.

- In 1803, Gian Francesco Malfatti claimed to prove that a certain arrangement of three circles would cover the maximum possible area inside a right triangle. However, to do so he made certain unwarranted assumptions about the configuration of the circles. It was shown in 1930 that circles in a different configuration could cover a greater area, and in 1967 that Malfatti's configuration was *never* optimal. See Malfatti circles.

- In 1806 André-Marie Ampère claimed to prove that a continuous function is differentiable at most points, but in 1872 Weierstrass gave an example of a continuous function that was not differentiable anywhere: The Weierstrass function.

- Dirichlet's theorem. In 1808 Legendre published an attempt at a proof of Dirichlet's theorem, but as Dupre pointed out in 1859 one of the lemmas used by Legendre is false. Dirichlet gave a complete proof in 1837.

- Uniform convergence. In his *Cours d'analyse* of 1821, Cauchy "proved" that if a sum of continuous functions converges pointwise, then its limit is also continuous. However, Abel observed three years later that this is not the case. For the conclusion to hold, "pointwise convergence" must be replaced with "uniform convergence".[1] There are many counterexamples. For example, a Fourier series of sine and cosine functions, all continuous, may converge to a discontinuous function such as a step function.

- Intersection theory. In 1848 Steiner claimed that the number of conics tangent to 5 given conics is $7776 = 6^5$, but later realized this was wrong. The correct number 3264 was found by Berner in 1865 and by de Jonquieres around 1859 and by Chasles in 1864 using his theory of characteristics. However these results, like many others in classical intersection theory, do not seem to have been given complete proofs until the work of Fulton and Macpherson in about 1978.

- Dirichlet's principle. This was used by Riemann in 1851, but Weierstrass found a counterexample to one version of this principle in 1870, and Hilbert stated and proved a correct version in 1900.

- In 1879, Alfred Kempe published a purported proof of the four-color map theorem, whose validity as a proof was accepted for eleven years before it was refuted. The proof did, however, suffice to show the weaker five-color map theorem. The four-color theorem was eventually proved in 1976.[2]

- Jordan curve theorem. There has been some controversy about whether Jordan's original proof of this in 1887 contains gaps. Oswald Veblen in 1905 claimed that Jordan's proof is incomplete, but in 2007 Hales said that the gaps are minor and that Jordan's proof is essentially complete.

- Wronskians. In 1887 Mansion claimed in his textbook that if a Wronskian of some functions vanishes everywhere then the functions are linearly dependent. In 1889 Peano pointed out the counterexample x^2 and $x|x|$. The result is correct if the functions are analytic.

- Vahlen (1891) published a purported example of an algebraic curve in 3-dimensional projective space that could not be defined as the zeros of 3 polynomials, but in 1941 Perron found 3 equations defining Vahlen's curve. In 1961 Kneser showed that any algebraic curve in projective 3-space can be given as the zeros of 3 polynomials.[3]

- In 1898 Miller published a paper incorrectly claiming to prove that the Mathieu group M_{24} does not exist, though in 1900 he pointed out that his proof was wrong.

- In 1905 Lebesgue tried to prove the (correct) result that a function implicitly defined by a Baire function is Baire, but his proof incorrectly assumed that the projection of a Borel set is Borel. Suslin pointed out the error and was inspired by it to define analytic sets as continuous images of Borel sets.

- Carmichael's totient function conjecture was stated as a theorem by Carmichael in 1907, but in 1922 he pointed out that his proof was incomplete. As of 2015 the problem is still open.

- Dehn's lemma. Dehn published an attempted proof in 1910, but Kneser found a gap in 1929. It was finally proven in 1956 by Christos Papakyriakopoulos.

- Italian school of algebraic geometry. Most gaps in proofs are caused either by a subtle technical oversight, or before the 20th century by a lack of precise definitions. A major exception to this is the Italian school of algebraic geometry in the first half of the 20th century, where lower standards of rigor gradually became acceptable. The result was that there are many papers in this area where the proofs are incomplete, or the theorems are not stated precisely. This list contains a few representative examples, where the result was not just incompletely proved but also hopelessly wrong.

- Perko pair, a pair of knots listed as distinct in tables for many years until Perko discovered in 1974 that they were the same. This gives a counterexample to a theorem claimed by Little in 1900 that the writhe of a reduced knot diagram is an invariant.

- Hilbert's sixteenth problem. Henri Dulac published a partial solution to this problem in 1923, but in about 1980 Écalle and Ilyashenko independently found a serious gap, and fixed it in about 1991.[4]

- Hilbert's twenty-first problem. In 1908 Plemelj claimed to have shown the existence a Fuchsian differential equations with any given monodromy group, but in 1989 Bolibruch discovered a counterexample.

- Groups of order 64. In 1930 Miller published a paper claiming that there are 294 groups of order 64. Hall and Senior showed in 1964 that the correct number is 267.

- Kurt Gödel proved in 1932 that the truth of a certain class of sentences of first-order arithmetic, known in the literature as $[\exists^* \forall^2 \exists^*, all, (0)]$, was decidable. That is, there was a method for deciding correctly whether any statement of that form was true. In the final sentence of that paper, he asserted that the same proof would work for the decidability of the larger class $[\exists^* \forall^2 \exists^*, all, (0)]_=$, which also includes formulas that contain an equality predicate. However, in the mid-1960s, Stål Aanderaa showed that Gödel's proof would *not* go through for the larger class, and in 1982 Warren Goldfarb showed that validity of formulas from the larger class was in fact undecidable.[5][6]

- Grunwald–Wang theorem. Wilhelm Grunwald published an incorrect proof in 1933 of an incorrect theorem, and Whaples later published another incorrect proof. Shianghao Wang found a counterexample in 1948 and published a corrected version of the theorem in 1950.

- In 1934 Severi claimed that the space of rational equivalence classes of cycles on an algebraic surface is finite-dimensional, but Mumford (1968) showed that this is false for surfaces of positive geometric genus.

- Littlewood–Richardson rule. Robinson published an incomplete proof in 1938, though the gaps were not noticed for many years. The first complete proofs were given by Schützenberger in 1977 and Thomas in 1974.

- Jacobian conjecture. Keller asked this as a question in 1939, and in the next few years there were several published incomplete proofs, including 3 by B. Segre, but Vitushkin found gaps in many of them. The Jacobian conjecture is (as of 2015) an open problem, and more incomplete proofs are regularly announced. Hyman Bass, Edwin H. Connell, and David Wright (1982) discuss the errors in some of these incomplete proofs.

- One of many examples from algebraic geometry in the first half of the 20th century: Severi (1946) claimed that that a degree-n surface in 3-dimensional projective space has at most $(n+2 \ 3)-4$ nodes, B. Segre pointed out that this was wrong; for example, for degree 6 the maximum number of nodes is 65, achieved by the Barth sextic, which is more than the maximum of 52 claimed by Severi.

- Rokhlin invariant. Rokhlin (1951) incorrectly claimed that the third stable stem of the homotopy groups of spheres is of order 12. In 1952 he discovered his error: it is in fact cyclic of order 24. The difference is crucial as it results in the existence of the Rokhlin invariant, a fundamental tool in the theory of 3- and 4-dimensional manifolds.

- Class numbers of imaginary quadratic fields. In 1952 Heegner published a solution to this problem. His paper was not accepted as a complete proof as it contained a gap, and the first complete proofs were given in about 1967 by Baker and Stark. In 1969 Stark showed how to fill the gap in Heegner's paper.

- Hilbert's sixteenth problem. In the 1950s, Evgenii Landis and Ivan Petrovsky published a purported solution, but it was shown wrong in the early 1960s.[4]

- Nielsen realization problem. Kravetz claimed to solve this in 1959 by first showing that Teichmuller space is negatively curved, but in 1974 Masur showed that it is not negatively curved. The Nielsen realization problem was finally solved in 1980 by Kerskhoff.

- Yamabe problem. Yamabe claimed a solution in 1960, but Trudinger discovered a gap in 1968, and a complete proof was not given until 1984.

- In 1961, Jan-Erik Roos published an incorrect theorem about the vanishing of the first derived functor of the inverse limit functor under certain general conditions.[7] However, over forty years later, Amnon Neeman constructed a counterexample.[8]

- Mordell conjecture over function fields. Manin published a proof in 1963, but Coleman (1990) found and corrected a gap in the proof.

- The Schur multiplier of the Mathieu group M_{22} is particularly notorious as it was miscalculated more than once: Burgoyne & Fong (1966) first claimed it had order 3, then in a 1968 correction claimed it had order 6; its order is in fact (currently believed to be) 12. This caused an error in the title of Janko's paper *A new finite simple group of order 86,775,570,046,077,562,880 which possesses M_{24} and the full covering group of M_{22} as subgroup* on J4: it does not have the full covering group as a subgroup, as the full covering group is larger than was realized at the time.

- Complex structures on the 6-sphere. In 1969 Alfred Adler published a paper in the American Journal of Mathematics claiming that the 6-sphere has no complex structure. His argument was incomplete, and this is (as of 2011) still a major open problem.

- In 1973 Britton published a 282 page attempted solution of Burnside's problem. In his proof he assumed the existence of a set of parameters satisfying some inequalities, but Adian pointed out that these inequalities were inconsistent. Novikov and Adian had previously found a correct solution around 1968.

- In 1975, Leitzel, Madan, and Queen incorrectly claimed that there are only 7 function fields over finite fields with genus >0 and class number 1, but in 2013 Stirpe found another; there are in fact exactly 8.

- Closed geodesics. In 1978 Wilhelm Klingenberg published a proof that smooth compact manifolds without boundary have infinitely many closed geodesics. His proof was controversial, and there is currently (as of 2011) no consensus on whether his proof is complete.

- Classification of finite simple groups. In 1983, Gorenstein announced that the proof of the classification had been completed, but he had been misinformed about the status of the proof of classification of quasithin groups, which had a serious gap in it. A complete proof for this case was published by Aschbacher and Smith in 2004.

- Kepler conjecture. Hsiang published an incomplete proof of this in 1993. Hales later published a proof (currently believed to be correct) depending on some very long computer calculations.

- Busemann–Petty problem. Zhang published two papers in the Annals of Mathematics in 1994 and 1999, in the first of which he proved that the Busemann–Petty problem in $\mathbf{R}^4$ has a negative solution, and in the second of which he proved that it has a positive solution.

- Algebraic stacks. The book Laumon & Moret-Bailly (2000) on algebraic stacks mistakenly claimed that morphisms of algebraic stacks induce morphisms of lisse-étale topoi. The results depending on this were repaired by Olsson (2007).

- Matroid bundles. In 2003 Biss published a paper in the Annals of Mathematics claiming to show that matroid bundles are equivalent to real vector bundles, but in 2009 published a correction pointing out a serious gap in the proof.

30.2   See also

- List of long mathematical proofs

30.3 References

[1] Porter, Roy (2003). *The Cambridge History of Science*. Cambridge University Press. p. 476. ISBN 0-521-57199-5.

[2] Thomas L. Saaty and Paul C. Kainen (1986). *The Four-Color Problem: Assaults and Conquest*. Dover Publications. ISBN 978-0-486-65092-0.

[3] http://mathoverflow.net/questions/35476

[4] Yulij Ilyashenko (2002). "Centennial History of Hilbert's 16th problem" (PDF). *Bulletin of the AMS* **39** (3): 301–354. doi:10.1090/s027 0979-02-00946-1.

[5] Boerger, Egon; Grädel, Erich; Gurevich, Yuri (1997). *The Classical Decision Problem*. Springer. p. 188. ISBN 3-540-42324-9.

[6] Goldfarb, Warren (1986). Feferman, Solomon, ed. *Kurt Gödel: Collected Works* **1**. Oxford University Press. pp. 229–231. ISBN 0-19-503964-5.

[7] Roos, Jan-Erik (1961). "Sur les foncteurs dérivés de lim. Applications.". *C. R. Acad. Sci. Paris* **252**: 3702–3704. MR 0132091.

[8] Neeman, Amnon (2002). "A counterexample to a 1961 "theorem" in homological algebra (with an appendix by P. Deligne)". *Inv. Math.* **148** (2): 397–420. doi:10.1007/s002220100197. MR 1906154.

- Bass, Hyman; Connell, Edwin H.; Wright, David (1982), "The Jacobian conjecture: reduction of degree and formal expansion of the inverse", *American Mathematical Society. Bulletin. New Series* **7** (2): 287–330, doi:10.1090/S0273-0979-1982-15032-7, ISBN 978-1-982150-32-7, MR 663785

- Burgoyne, N.; Fong, Paul (1966), "The Schur multipliers of the Mathieu groups", *Nagoya Mathematical Journal* **27**: 733–745, ISSN 0027-7630, MR 0197542

- Coleman, Robert F. (1990), "Manin's proof of the Mordell conjecture over function fields", *L'Enseignement Mathématique. Revue Internationale. IIe Série* **36** (3): 393–427, ISSN 0013-8584, MR 1096426

- Laumon, Gérard; Moret-Bailly, Laurent (2000), *Champs algébriques*, Ergebnisse der Mathematik und ihrer Grenzgebiete. 3. Folge. A Series of Modern Surveys in Mathematics [Results in Mathematics and Related Areas. 3rd Series. A Series of Modern Surveys in Mathematics] **39**, Berlin, New York: Springer-Verlag, ISBN 978-3-540-65761-3, MR 1771927

- Lecat, Maurice (1935), *Erreurs de mathématiciens des origines à nos jours*, Bruxelles - Louvain: Librairie Castaigne - Ém. Desbarax

- Mumford, David (1968), "Rational equivalence of 0-cycles on surfaces", *Journal of Mathematics of Kyoto University* **9**: 195–204, ISSN 0023-608X, MR 0249428

- Olsson, Martin (2007), "Sheaves on Artin stacks", *Journal für die reine und angewandte Mathematik* **603**: 55–112, doi:10.1515/CRELLE.2007.012, ISSN 0075-4102, MR 2312554

- Rohlin, V. A. (1951), "Classification of mappings of an (n+3)-dimensional sphere into an n-dimensional one", *Doklady Akad. Nauk SSSR (N.S.)* **81**: 19–22, MR 0046043

- Severi, Francesco (1946), "Sul massimo numero di nodi di una superficie di dato ordine dello spazio ordinario o di una forma di un operspazio", *Annali di Matematica Pura ed Applicata. Serie Quarta* **25**: 1–41, doi:10.1007/bf02418077, ISSN 0003-4622

- Vahlen, K. T. (1891), "Bemerkung zur vollställndigen Darstellung algebraischer Raumkurven", *J. Reine Angew. Math.* **108**: 346–347

30.4 External links

- David Mumford email about the errors of the Italian algebraic geometry school under Severi

30.4.1 MathOverflow questions

- Ilya Nikokoshev, Most interesting mathematics mistake?

- Kevin Buzzard what mistakes did the Italian algebraic geometers actually make?

- Will Jagy, Widely accepted mathematical results that were later shown wrong?

- John Stillwell, What are some correct results discovered with incorrect (or no) proofs?

Chapter 31

List of long mathematical proofs

This is a list of unusually long mathematical proofs.

As of 2011, the longest mathematical proof, measured by number of published journal pages, is the classification of finite simple groups with well over 10000 pages. There are several proofs that would be far longer than this if the details of the computer calculations they depend on were published in full.

31.1 Long proofs

The length of unusually long proofs has increased with time. As a rough rule of thumb, 100 pages in 1900, or 200 pages in 1950, or 500 pages in 2000 is unusually long for a proof.

- 1799 The Abel–Ruffini theorem was nearly proved by Paolo Ruffini, but his proof, spanning 500 pages, was mostly ignored and later, in 1824, Niels Henrik Abel published a proof that required just six pages

- 1890 Killing's classification of simple complex Lie algebras, including his discovery of the exceptional Lie algebras, took 180 pages in 4 papers.

- 1894 The ruler-and-compass construction of a polygon of 65537 sides by Johann Gustav Hermes took over 200 pages.

- 1905 Emanuel Lasker's original proof of the Lasker–Noether theorem took 98 pages, but has since been simplified: modern proofs are less than a page long.

- 1963 Odd order theorem This was 255 pages long, which at the time was over 10 times as long as what had previously been considered a long paper in group theory.

- 1964 Resolution of singularities Hironaka's original proof was 216 pages long; it has since been simplified considerably down to about 10 or 20 pages.

- 1966 Abyhankar's proof of resolution of singularities for 3-folds in characteristic greater than 6 covered about 500 pages in several papers. (In 2009 Cutkosky simplified this to about 40 pages.)

- 1966 Discrete series representations of Lie groups. Harish-Chandra's construction of these involved a long series of papers totaling around 500 pages. His later work on the Plancherel theorem for semisimple groups added another 150 pages to these.

- 1968 the Novikov-Adian proof solving Burnside's problem on finitely generated infinite groups with finite exponents negatively. The three-part original paper is more than 300 pages long. (Britton later published a 282 page paper attempting to solve the problem, but his paper contained a serious gap.)

- 1960–1970 Fondements de la Géometrie Algébrique, Éléments de géométrie algébrique and Séminaire de géométrie algébrique. Grothendieck's work on the foundations of algebraic geometry covers many thousands of pages. Although this is not a proof of a single theorem, there are several theorems in it whose proofs depend on hundreds of earlier pages.

- 1974 N-group theorem Thompson's classification of N-groups used 6 papers totaling about 400 pages, but also used earlier results of his such as the odd order theorem, which bring to total length up to more than 700 pages.

- 1974 Ramanujan conjecture and the Weil conjectures. While Deligne's final paper proving these was "only" about 30 pages long, it depended on background results in algebraic geometry and étale cohomology that Deligne estimated to be about 2000 pages long.

- 1974 4-color theorem. Appel and Haken's proof of this took 139 pages, and also depended on long computer calculations.

- 1974 The Gorenstein–Harada theorem classifying finite groups of sectional 2-rank at most 4 was 464 pages long.

- 1976 Eisenstein series Langlands's proof of the functional equation for Eisenstein series was 337 pages long.

- 1983 Trichotomy theorem Gorenstein and Lyons's proof for the case of rank at least 4 was 731 pages long, and Aschbacher's proof of the rank 3 case adds another 159 pages, for a total of 890 pages.

- 1983 Selberg trace formula Hejhal's proof of a general form of the Selberg trace formula consisted of 2 volumes with a total length of 1322 pages.

- Arthur–Selberg trace formula. Arthur's proofs of the various versions of this cover several hundred pages spread over many papers.

- 2000 Almgren's regularity theorem Almgren's proof was 955 pages long.

- 2000 Lafforgue's theorem on the Langlands conjecture for the general linear group over function fields. Laurent Lafforgue's proof of this was about 600 pages long, not counting many pages of background results.

- 2003 Poincaré conjecture, Geometrization theorem, Geometrization conjecture. Perelman's original proofs of the Poincaré conjecture and the Geometrization conjecture were not lengthy, but were rather sketchy. Several other mathematicians have published proofs with the details filled in, which come to several hundred pages.

- 2004 Quasithin groups The classification of the simple quasithin groups by Aschbacher and Smith was 1221 pages long, one of the longest single papers ever written.

- 2004 Classification of finite simple groups. The proof of this is spread out over hundreds of journal articles which makes it hard to estimate its total length, which is probably around 10000 to 20000 pages.

- 2004 Robertson–Seymour theorem. The proof takes about 500 pages spread over about 20 papers.

- 2005 Kepler conjecture Hales's proof of this involves several hundred pages of published arguments, together with several gigabytes of computer calculations.

- 2006 the strong perfect graph theorem, by Maria Chudnovsky, Neil Robertson, Paul Seymour, and Robin Thomas. 180 pages in the Annals of Mathematics.

- 2012 Inter-universal Teichmüller theory Mochizuki's work on this (which is still being verified) covers many hundreds of pages spread over several long papers.

31.2 Long computer calculations

There are many mathematical theorems that have been checked by long computer calculations. If these were written out as proofs many would be far longer than most of the proofs above. There is not really a clear distinction between computer calculations and proofs, as several of the proofs above, such as the 4-color theorem and the Kepler conjecture, use long computer calculations as well as many pages of mathematical argument. For the computer calculations in this section, the mathematical arguments are only a few pages long, and the length is due to long but routine calculations. Some typical examples of such theorems include:

- Several proofs of the existence of sporadic simple groups, such as the Lyons group, originally used computer calculations with large matrices or with permutations on billions of symbols. In most cases, such as the baby monster group, the computer proofs were later replaced by shorter proofs avoiding computer calculations. Similarly the calculation of the maximal subgroups of the larger sporadic groups uses a lot of computer calculations.

- 2004 Verification of the Riemann hypothesis for the first 10^{13} zeros of the Riemann zeta function.

- 2007 Verification that Checkers is a draw.

- 2008 Proofs that various Mersenne numbers with around ten million digits are prime.

- Calculations of large numbers of digits of π.

- 2010 Showing that Rubik's Cube can be solved in 20 moves.

- 2012 Showing that Sudoku needs at least 17 clues .

- 2013 Ternary Goldbach conjecture: Every odd number greater than 5 can be expressed as the sum of three primes.

- 2014 Proof of Erdős discrepancy conjecture for particular case C=2: every ±1-sequence of the length 1161 has a discrepancy at least 3, original proof generated by a SAT solver had a size of 13 gigabytes, it has been reduced later to 850 megabytes.

31.3 Long proofs in mathematical logic

Main article: Gödel's speed-up theorem

Kurt Gödel showed how to find explicit examples of statements in formal systems that are provable in that system but whose shortest proof is absurdly long. For example, the statement:

"This statement cannot be proved in Peano arithmetic in less than a googolplex symbols"

is provable in Peano arithmetic but the shortest proof has at least a googolplex symbols. It has a short proof in a more powerful system: in fact it is easily provable in Peano arithmetic together with the statement that Peano arithmetic is consistent (which cannot be proved in Peano arithmetic by Gödel's incompleteness theorem).

In this argument, Peano arithmetic can be replaced by any more powerful consistent system, and a googolplex can be replaced by any number that can be described concisely in the system.

Harvey Friedman found some explicit natural examples of this phenomenon, giving some explicit statements in Peano arithmetic and other formal systems whose shortest proofs are ridiculously long (Smoryński 1982). For example, the statement that

"there is an integer n such that if there is a sequence of rooted trees T_1, T_2, ..., Tn such that Tk has at most $k+10$ vertices, then some tree can be homeomorphically embedded in a later one"

is provable in Peano arithmetic, but the shortest proof has length at least $A(1000)$, where $A(0)=1$ and $A(n+1)=2^{A(n)}$. The statement is a special case of Kruskal's theorem and has a short proof in second order arithmetic.

31.4   See also

- List of incomplete proofs

31.5   References

- Krantz, Steven G. (2011), *The proof is in the pudding. The changing nature of mathematical proof* (PDF), Berlin, New York: Springer-Verlag, ISBN 978-0-387-48908-7, MR 2789493

- Smoryński, C. (1982), "The varieties of arboreal experience", *Math. Intelligencer* **4** (4): 182–189, doi:10.1007/bf03023553, MR 0685558

Chapter 32

List of mathematical proofs

A list of articles with mathematical proofs:

32.1 Theorems of which articles are primarily devoted to proving them

See also: Category:Article proofs

- Bertrand's postulate and a proof
- Estimation of covariance matrices
- Fermat's little theorem and some proofs
- Gödel's completeness theorem and its original proof
- Mathematical induction and a proof
- Proof that 0.999... equals 1
- Proof that 22/7 exceeds π
- Proof that e is irrational
- Proof that π is irrational
- Proof that the sum of the reciprocals of the primes diverges

32.2 Articles devoted to theorems of which a (sketch of a) proof is given

See also: Category:Articles containing proofs

- Banach fixed point theorem
- Banach–Tarski paradox
- Basel problem
- Bolzano–Weierstrass theorem

183

- Brouwer fixed point theorem
- Buckingham π theorem (proof in progress)
- Burnside's lemma
- Cantor's theorem
- Cantor–Bernstein–Schroeder theorem
- Cayley's formula
- Cayley's theorem
- Clique problem (to do)
- Compactness theorem (very compact proof)
- Erdős–Ko–Rado theorem
- Euler's formula
- Euler's four-square identity
- Euler's theorem
- Five color theorem
- Five lemma
- Fundamental theorem of arithmetic
- Gauss–Markov theorem (brief pointer to proof)
- Gödel's incompleteness theorem
    - Gödel's first incompleteness theorem
    - Gödel's second incompleteness theorem
- Goodstein's theorem
- Green's theorem (to do)
    - Green's theorem when D is a simple region
- Heine–Borel theorem
- Intermediate value theorem
- Itō's lemma
- König's lemma
- König's theorem (set theory)
- König's theorem (graph theory)
- Lagrange's theorem
- Liouville's theorem (brief pointer to proof)
- Markov's inequality (proof of a generalization)
- Mean value theorem

- Multivariate normal distribution (to do)
- Holomorphic functions are analytic
- Pythagorean theorem
- Quadratic equation
- Quotient rule
- Ramsey's theorem
- Rao–Blackwell theorem
- Rice's theorem
- Rolle's theorem
- Splitting lemma
- squeeze theorem
- Sum rule in differentiation
- Sum rule in integration
- Sylow theorems
- Transcendence of e and π (as corollaries of Lindemann–Weierstrass)
- Tychonoff's theorem (to do)
- Ultrafilter lemma
- Ultraparallel theorem
- Urysohn's lemma
- Van der Waerden's theorem
- Wilson's theorem
- Zorn's lemma

32.3   Articles devoted to algorithms in which their correctness is proven

- Bellman–Ford algorithm (to do)
- Euclidean algorithm
- Kruskal's algorithm
- Gale–Shapley algorithm
- Prim's algorithm
- Shor's algorithm (incomplete)

32.4   Articles where example statements are proven

See also: Category:Articles containing proofs

- Basis (linear algebra)
- Burrows–Abadi–Needham logic
- Direct proof
- Generating a vector space
- Linear independence
- Polynomial
- Proof
- Pumping lemma
- Simpson's rule

32.5   Other articles containing proofs

See also: Category:Articles containing proofs

- Addition in N
    - associativity of addition in N
    - commutativity of addition in N
    - uniqueness of addition in N
- Algorithmic information theory
- Boolean ring
    - commutativity of a boolean ring
- Boolean satisfiability problem
    - NP-completeness of the Boolean satisfiability problem
- Cantor's diagonal argument
    - set is smaller than its power set
    - uncountability of the real numbers
- Cantor's first uncountability proof
    - uncountability of the real numbers
- Combinatorics
- Combinatory logic
- Co-NP

- Coset
- Countable
  - countability of a subset of a countable set (to do)
- Angle of parallelism
- Galois group
  - Fundamental theorem of Galois theory (to do)
- Gödel number
  - Gödel's incompleteness theorem
- Group (mathematics)
- Halting problem
  - insolubility of the halting problem
- Harmonic series (mathematics)
  - divergence of the (standard) harmonic series
- Highly composite number
- Area of hyperbolic sector, basis of hyperbolic angle
- Infinite series
  - convergence of the geometric series with first term 1 and ratio 1/2
- Integer partition
- Irrational number
  - irrationality of $\log_2 3$
  - irrationality of the square root of 2
- Limit point
- Mathematical induction
  - sum identity
- Power rule
  - differential of x^n
- Product and Quotiont Rules
- Derivation of Product and Quotient rules for differentiating.
- Prime number
  - Infinitude of the prime numbers
- Primitive recursive function
- Principle of bivalence
  - no propositions are neither true nor false in intuitionistic logic

- Recursion
- Relational algebra (to do)
- Solvable group
- Square root of 2
- Tetris
- Algebra of sets
    - idempotent laws for set union and intersection

32.6   Articles which mention dependencies of theorems

- Cauchy's integral formula
- Cauchy integral theorem
- Computational geometry
- Fundamental theorem of algebra
- Lambda calculus
- Invariance of domain
- Minkowski inequality
- Nash embedding theorem
- Open mapping theorem (functional analysis)
- Product topology
- Riemann integral
- Time hierarchy theorem
    - Deterministic time hierarchy theorem

32.7   Articles giving mathematical proofs within a physical model

- No cloning theorem
- Torque

32.8   See also

- Gödel's ontological proof
- Invalid proof
- List of theorems
- List of incomplete proofs
- List of long proofs

Chapter 33

Proof by intimidation

Proof by intimidation (or argumentum verbosium) is a jocular phrase used mainly in mathematics to refer to a style of presenting a purported mathematical proof by giving an argument loaded with jargon and appeal to obscure results, so that the audience is simply obliged to accept it, lest they have to admit their ignorance and lack of understanding.[1]

The phrase is also used when the author is an authority in his field presenting his proof to people who respect *a priori* his insistence that the proof is valid or when the author claims that his statement is true because it is trivial or because he simply says so. Usage of this phrase is for the most part in good humour, though it also appears in serious criticism.[2]

"Proof by intimidation" is also cited by critics of junk science to describe cases in which scientific evidence is thrown aside in favour of a litany of tragic individual cases presented to the public by articulate advocates who pose as experts in their field.[3]

Gian-Carlo Rota claimed in a memoir that the expression "proof by intimidation" was coined by Mark Kac to describe a technique used by William Feller in his lectures.[4]

33.1 See also

- *Ad nauseam*
- Argument from authority
- Chewbacca defense
- Gish Gallop
- Handwaving
- Obscurantism
- Sophism

33.2 References

[1] Michael H. F. Wilkinson. "Cogno-Intellectualism, Rhetorical Logic, and the Craske-Trump Theorem". *Annals of Improbable Research* **6** (5): 15–16. Retrieved 2008-02-22.

[2] Tony Hey (1999). "Richard Feynman and computation". *Contemporary Physics* **40** (4): 257–265. doi:10.1080/001075199181459. Retrieved 2008-02-22.

[3] Marjorie K. Jeffcoat (July 2003). "Junk science: Appearances can be deceiving". *Journal of the American Dental Association* **134** (7): 802–803. doi:10.14219/jada.archive.2003.0268. PMID 12892436. Retrieved 2008-02-22.

[4] *He took umbrage when someone interrupted his lecturing by pointing out some glaring mistake. He became red in the face and raised his voice, often to full shouting range. It was reported that on occasion he had asked the objector to leave the classroom. The expression "proof by intimidation" was coined after Feller's lectures (by Mark Kac). During a Feller lecture, the hearer was made to feel privy to some wondrous secret, one that often vanished by magic as he walked out of the classroom at the end of the period. Like many great teachers, Feller was a bit of a con man.* Proof by intimidation was also referenced in xkcd: http://xkcd.com/982/. Gian-Carlo Rota (1996). *Indiscrete Thoughts.* Boston: Birkhäuser. ISBN 0-8176-3866-0.

Chapter 34

Termination analysis

In computer science, a **termination analysis** is program analysis which attempts to determine whether the evaluation of a given program will definitely terminate. Because the halting problem is undecidable, termination analysis cannot be total. The aim is to find the answer "program does terminate" (or "program does not terminate") whenever this is possible. Without success the algorithm (or human) working on the termination analysis may answer with "maybe" or continue working infinitely long.

34.1   Termination proof

A *termination proof* is a type of mathematical proof that plays a critical role in formal verification because total correctness of an algorithm depends on termination.

A simple, general method for constructing termination proofs involves associating a **measure** with each step of an algorithm. The measure is taken from the domain of a well-founded relation, such as from the ordinal numbers. If the measure "decreases" according to the relation along every possible step of the algorithm, it must terminate, because there are no infinite descending chains with respect to a well-founded relation.

Some types of termination analysis can automatically generate or imply the existence of a termination proof.

34.2   Example

An example of a programming language construct which may or may not terminate is a loop, as they can be run repeatedly. Loops implemented using a counter variable as typically found in data processing algorithms will usually terminate, demonstrated by the pseudocode example below:

i := 0 **loop** until i = SIZE_OF_DATA process_data(data[i])) //process the data chunk at position i i := i + 1 //move to the next chunk of data to be processed

If the value of *SIZE_OF_DATA* is non-negative, fixed and finite, the loop will eventually terminate, assuming *process_data* terminates too.

Some loops can be shown to always terminate or never terminate, through human inspection. For example, even a non-programmer should see that, in theory, the following never stops (but it may halt on physical machines due to arithmetic overflow):

i := 1 **loop** until i = 0 i := i + 1

In termination analysis one may also try to determine the termination behaviour of some program depending on some unknown input. The following example illustrates this problem.

i := 1 **loop** until i = UNKNOWN i := i + 1

191

Here the loop condition is defined using some value UNKNOWN, where the value of UNKNOWN is not known (e.g. defined by the user's input when the program is executed). Here the termination analysis must take into account all possible values of UNKNOWN and find out that in the possible case of UNKNOWN = 0 (as in the original example) the termination cannot be shown.

There is, however, no general procedure for determining whether an expression involving looping instructions will halt, even when humans are tasked with the inspection. The theoretical reason for this is the undecidability of the Halting Problem: there cannot exist some algorithm which determines whether any given program stops after finitely many computation steps.

In practice one fails to show termination (or non-termination) because every algorithm works with a finite set of methods being able to extract relevant information out of a given program. A method might look at how variables change with respect to some loop condition (possibly showing termination for that loop), other methods might try to transform the program's calculation to some mathematical construct and work on that, possibly getting information about the termination behaviour out of some properties of this mathematical model. But because each method is only able to "see" some specific reasons for (non)termination, even through combination of such methods one cannot cover all possible reasons for (non)termination.

Recursive functions and loops are equivalent in expression; any expression involving loops can be written using recursion, and vice versa. Thus the termination of recursive expressions is also undecidable in general. Most recursive expressions found in common usage (i.e. not pathological) can be shown to terminate through various means, usually depending on the definition of the expression itself. As an example, the function argument in the recursive expression for the factorial function below will always decrease by 1; from the well-ordering property on natural numbers, the argument will eventually reach 1 and the recursion will terminate.

function factorial (argument **as** natural number) **if** argument = 0 **or** argument = 1 **return** 1 **otherwise return** argument * factorial(argument - 1)

34.3   Dependent types

Termination check is very important in dependently typed programming language and theorem proving systems like Coq and Agda. These systems use Curry-Howard isomorphism between programs and proofs. Proofs over inductively defined data types were traditionally described using induction and recursion principles which are in fact, primitive recursion. However, it was found later, that describing a program via a recursively defined function with pattern matching is more natural way of proving than using induction principle directly. Unfortunately, allowing arbitrary, including non terminating definitions, leads to possibility of logical inconsistencies in type theories. That's why Agda and Coq have termination checkers built-in.

34.3.1   Sized types

One of the approaches to termination checking in dependently typed programming languages are sized types. The main idea is to annotate the types over which we can recurse with size annotations and allow recursive calls only on smaller arguments. Sized types are implemented in Agda as a syntactic extension.

34.4   Current Research

There are several research teams that work on new methods that can show (non)termination. Many researchers include these methods into programs[1] that try to analyze the termination behavior automatically (so without human interaction). An on-going aspect of research is to allow the existing methods to be used to analyze termination behavior of programs written in "real world" programming languages. For declarative languages like Haskell, Mercury and Prolog, many results exist[2][3][4] (mainly because of the strong mathematical background of these languages). The research community also works on new methods to analyze termination behavior of programs written in imperative languages like C and Java.

Because of the undecidability of the Halting Problem research in this field cannot reach completeness. One can always think of new methods that find new (complicated) reasons for termination.

34.5   See also

- Complexity analysis — the problem of estimating the time needed to terminate

- Loop variant

- Total functional programming — a programming paradigm that restricts the range of programs to those that are provably terminating

- Walther recursion

34.6   References

[1] Tools at termination-portal.org

[2] Giesl, J. and Swiderski, S. and Schneider-Kamp, P. and Thiemann, R. Pfenning, F., ed. *Automated Termination Analysis for Haskell: From Term Rewriting to Programming Languages (invited lecture)* (POSTSCRIPT). Term Rewriting and Applications, 17th Int. Conf., RTA-06. LNCS **4098**. pp. 297–312.

[3] Compiler options for termination analysis in Mercury

[4] http://verify.rwth-aachen.de/giesl/papers/lopstr07-distribute.pdf

Research papers on automated program termination analysis include:

- Christoph Walther (1988). "Argument-Bounded Algorithms as a Basis for Automated Termination Proofs". *Proc. 9th Conference on Automated Deduction*. LNAI **310**. Springer. pp. 602–621.

- Christoph Walther (1991). "On Proving the Termination of Algorithms by Machine" (PDF). *Artificial Intelligence* **70** (1).

- Xi, Hongwei (1998). "Towards Automated Termination Proofs through *Freezing*" (PDF). In Tobias Nipkow. *Rewriting Techniques and Applications, 9th Int. Conf., RTA-98*. LNCS **1379**. Springer. pp. 271–285.

- Jürgen Giesl; Christoph Walther; Jürgen Brauburger (1998). "Termination Analysis for Functional Programs". In W. Bibel; P. Schmitt. *Automated Deduction - A Basis for Applications* (POSTSCRIPT) **3**. Dordrecht: Kluwer Academic Publishers. pp. 135–164.

- Christoph Walther (2000). "Criteria for Termination". In S. Hölldobler. *Intellectics and Computational Logic* (POSTSCRIPT). Dordrecht: Kluwer Academic Publishers. pp. 361–386.

- Christoph Walther; Stephan Schweitzer (2005). "Automated Termination Analysis for Incompletely Defined Programs" (PDF). In Franz Baader; Andrei Voronkov. *Proc. 11th Int. Conf. on Logic for Programming, Artificial Intelligence and Reasoning (LPAR)*. LNAI **3452**. Springer. pp. 332–346.

- Adam Koprowski; Johannes Waldmann (2008). "Arctic Termination ...Below Zero". In Andrei Voronkov. *Rewriting Techniques and Applications, 19th Int. Conf., RTA-08* (PDF). Lecture Notes in Computer Science **5117**. Springer. pp. 202–216. ISBN 978-3-540-70588-8.

System descriptions of automated termination analysis tools include:

- Giesl, J. (1995). "Generating Polynomial Orderings for Termination Proofs (system description)". In Hsiang, Jieh. *Rewriting Techniques and Applications, 6th Int. Conf., RTA-95* (POSTSCRIPT). LNCS **914**. Springer. pp. 426–431.

- Ohlebusch, E.; Claves, C.; Marché, C. (2000). "TALP: A Tool for the Termination Analysis of Logic Programs (system description)". In Bachmair, Leo. *Rewriting Techniques and Applications, 11th Int. Conf., RTA-00* (COMPRESSED POSTSCRIPT). LNCS **1833**. Springer. pp. 270–273.

- Hirokawa, N.; Middeldorp, A. (2003). "Tsukuba Termination Tool (system description)". In Nieuwenhuis, R. *Rewriting Techniques and Applications, 14th Int. Conf., RTA-03* (PDF). LNCS **2706**. Springer. pp. 311–320.

- Giesl, J.; Thiemann, R.; Schneider-Kamp, P.; Falke, S. (2004). "Automated Termination Proofs with APro VE (system description)". In van Oostrom, V. *Rewriting Techniques and Applications, 15th Int. Conf., RTA-04* (PDF). LNCS **3091**. Springer. pp. 210–220. ISBN 3-540-22153-0.

- Hirokawa, N.; Middeldorp, A. (2005). "Tyrolean Termination Tool (system description)". In Giesl, J. *Term Rewriting and Applications, 16th Int. Conf., RTA-05*. LNCS **3467**. Springer. pp. 175–184. ISBN 978-3-540-25596-3.

- Koprowski, A. (2006). "TPA: Termination Proved Automatically (system description)". In Pfenning, F. *Term Rewriting and Applications, 17th Int. Conf., RTA-06*. LNCS **4098**. Springer. pp. 257–266.

- Marché, C.; Zantema, H. (2007). "The Termination Competition (system description)". In Baader, F. *Term Rewriting and Applications, 18th Int. Conf., RTA-07* (PDF). LNCS **4533**. Springer. pp. 303–313.

34.7   External links

- Termination Analysis of Higher-Order Functional Programs

- Termination Tools mailing list

- Termination Competition — see Marché, Zantema (2007) for a description

- Termination Portal

Chapter 35

Mathematical beauty

Mathematical beauty describes the notion that some mathematicians may derive aesthetic pleasure from their work, and from mathematics in general. They express this pleasure by describing mathematics (or, at least, some aspect of mathematics) as *beautiful*. Mathematicians describe mathematics as an art form or, at a minimum, as a creative activity. Comparisons are often made with music and poetry.

Bertrand Russell expressed his sense of mathematical beauty in these words:

> Mathematics, rightly viewed, possesses not only truth, but supreme beauty — a beauty cold and austere, like that of sculpture, without appeal to any part of our weaker nature, without the gorgeous trappings of painting or music, yet sublimely pure, and capable of a stern perfection such as only the greatest art can show. The true spirit of delight, the exaltation, the sense of being more than Man, which is the touchstone of the highest excellence, is to be found in mathematics as surely as poetry.[1]

Paul Erdős expressed his views on the ineffability of mathematics when he said, "Why are numbers beautiful? It's like asking why is Beethoven's Ninth Symphony beautiful. If you don't see why, someone can't tell you. I *know* numbers are beautiful. If they aren't beautiful, nothing is".[2]

35.1 Beauty in method

Mathematicians describe an especially pleasing method of proof as *elegant*. Depending on context, this may mean:

- A proof that uses a minimum of additional assumptions or previous results.

- A proof that is unusually succinct.

- A proof that derives a result in a surprising way (e.g., from an apparently unrelated theorem or collection of theorems.)

- A proof that is based on new and original insights.

- A method of proof that can be easily generalized to solve a family of similar problems.

In the search for an elegant proof, mathematicians often look for different independent ways to prove a result—the first proof that is found may not be the best. The theorem for which the greatest number of different proofs have been discovered is possibly the Pythagorean theorem, with hundreds of proofs having been published.[3] Another theorem that has been proved in many different ways is the theorem of quadratic reciprocity—Carl Friedrich Gauss alone published eight different proofs of this theorem.

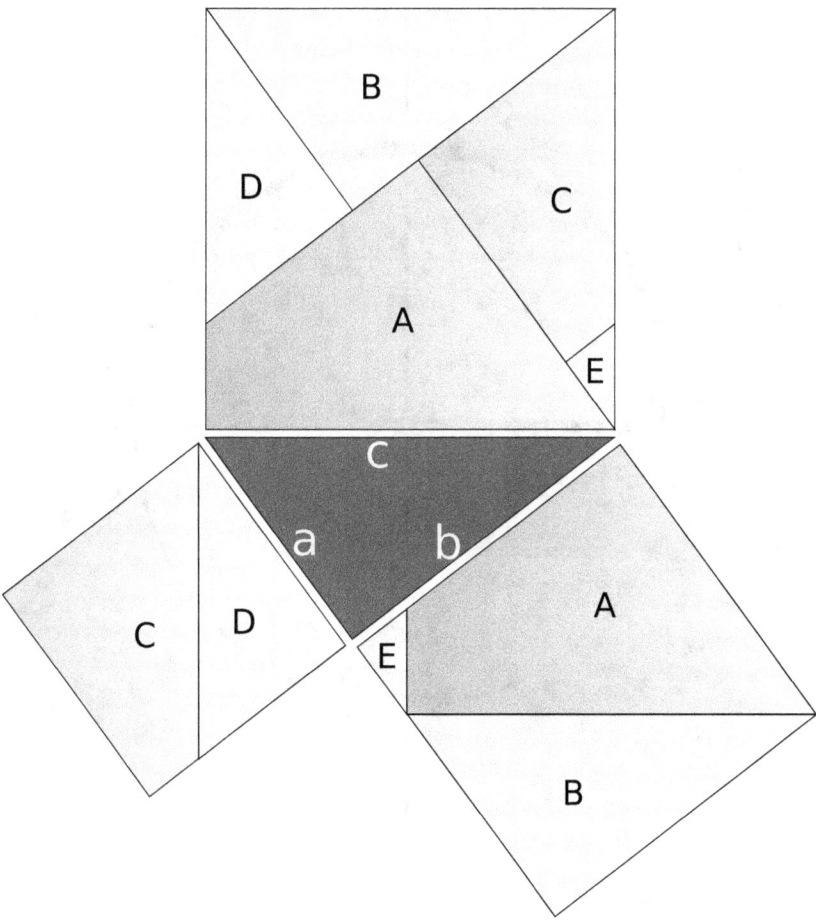

An example of "beauty in method"—a simple and elegant proof of the Pythagorean theorem.

Conversely, results that are logically correct but involve laborious calculations, over-elaborate methods, very conventional approaches, or that rely on a large number of particularly powerful axioms or previous results are not usually considered to be elegant, and may be called *ugly* or *clumsy*.

35.2   Beauty in results

Some mathematicians[4] see beauty in mathematical results that establish connections between two areas of mathematics that at first sight appear to be unrelated. These results are often described as *deep*.

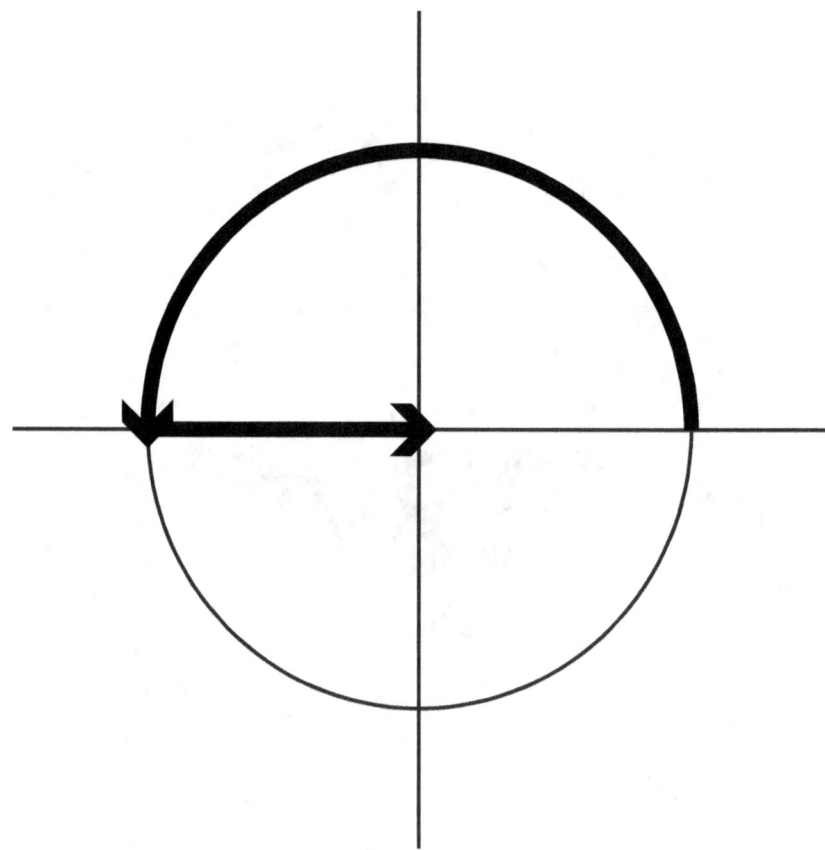

Starting at $e^0 = 1$, *travelling at the velocity* i *relative to one's position for the length of time* π, *and adding 1, one arrives at 0. (The diagram is an Argand diagram)*

While it is difficult to find universal agreement on whether a result is deep, some examples are often cited. One is Euler's identity:[5]

$$e^{i\pi} + 1 = 0.$$

This is a special case of Euler's formula, which the physicist Richard Feynman called "our jewel" and "the most remarkable formula in mathematics".[6] Modern examples include the modularity theorem, which establishes an important connection between elliptic curves and modular forms (work on which led to the awarding of the Wolf Prize to Andrew Wiles and Robert Langlands), and "monstrous moonshine", which connects the Monster group to modular functions via string theory for which Richard Borcherds was awarded the Fields Medal.

Other examples of deep results include unexpected insights into mathematical structures. For example, Gauss's Theorema Egregium is a deep theorem which relates a local phenomenon (curvature) to a global phenomenon (area) in a surprising way. In particular, the area of a triangle on a curved surface is proportional to the excess of the triangle and the proportionality is curvature. Another example is the fundamental theorem of calculus (and its vector versions including Green's theorem and Stokes' theorem).

The opposite of *deep* is *trivial*. A trivial theorem may be a result that can be derived in an obvious and straightforward way from other known results, or which applies only to a specific set of particular objects such as the empty set. Sometimes, however, a statement of a theorem can be original enough to be considered deep, even though its proof is fairly obvious.

In his *A Mathematician's Apology*, Hardy suggests that a beautiful proof or result possesses "inevitability", "unexpectedness", and "economy".[7]

Rota, however, disagrees with unexpectedness as a sufficient condition for beauty and proposes a counterexample:

> A great many theorems of mathematics, when first published, appear to be surprising; thus for example some twenty years ago [from 1977] the proof of the existence of non-equivalent differentiable structures on spheres of high dimension was thought to be surprising, but it did not occur to anyone to call such a fact beautiful, then or now.[8]

Perhaps ironically, Monastyrsky writes:

> It is very difficult to find an analogous invention in the past to Milnor's beautiful construction of the different differential structures on the seven-dimensional sphere....The original proof of Milnor was not very constructive, but later E. Briscom showed that these differential structures can be described in an extremely explicit and beautiful form.[9]

This disagreement illustrates both the subjective nature of mathematical beauty and its connection with mathematical results: in this case, not only the existence of exotic spheres, but also a particular realization of them.

35.3   Beauty in experience

Interest in pure mathematics separate from empirical study has been part of the experience of various civilizations, including that of the Ancient Greeks, who "did mathematics for the beauty of it."[10] Mathematical beauty can also be experienced outside the confines of pure mathematics. For example, the aesthetic pleasure that mathematical physicists tend to experience in Einstein's theory of general relativity has been attributed (by Paul Dirac, among others) to its "great mathematical beauty."[11]

Some degree of delight in the manipulation of numbers and symbols is probably required to engage in any mathematics. Given the utility of mathematics in science and engineering, it is likely that any technological society will actively cultivate these aesthetics, certainly in its philosophy of science if nowhere else.

The most intense experience of mathematical beauty for most mathematicians comes from actively engaging in mathematics. It is very difficult to enjoy or appreciate mathematics in a purely passive way—in mathematics there is no real analogy of the role of the spectator, audience, or viewer.[12] Bertrand Russell referred to the *austere beauty* of mathematics.

35.4   Beauty and philosophy

Some mathematicians are of the opinion that the doing of mathematics is closer to discovery than invention, for example:

> There is no scientific discoverer, no poet, no painter, no musician, who will not tell you that he found ready made his discovery or poem or picture – that it came to him from outside, and that he did not consciously create it from within.
> —William Kingdon Clifford, from a lecture to the Royal Institution titled "Some of the conditions of mental development"

There is a certain "cold and austere" beauty in this compound of five cubes

These mathematicians believe that the detailed and precise results of mathematics may be reasonably taken to be true without any dependence on the universe in which we live. For example, they would argue that the theory of the natural numbers is fundamentally valid, in a way that does not require any specific context. Some mathematicians have extrapolated this viewpoint that mathematical beauty is truth further, in some cases becoming mysticism.

Pythagorean mathematicians believed in the literal reality of numbers. The discovery of the existence of irrational numbers was a shock to them, since they considered the existence of numbers not expressible as the ratio of two natural numbers to be a flaw in nature (the Pythagorean world view did not contemplate the limits of infinite sequences of ratios of natural numbers—the modern notion of a real number). From a modern perspective, their mystical approach to numbers may be viewed as numerology.

In Plato's philosophy there were two worlds, the physical one in which we live and another abstract world which contained unchanging truth, including mathematics. He believed that the physical world was a mere reflection of the more perfect abstract world.

Hungarian mathematician Paul Erdős[13] spoke of an imaginary book, in which God has written down all the most beautiful mathematical proofs. When Erdős wanted to express particular appreciation of a proof, he would exclaim "This one's from The Book!" This viewpoint expresses the idea that mathematics, as the intrinsically true foundation on which the laws of our universe are built, is a natural candidate for what has been personified as God by different religious believers.

Twentieth-century French philosopher Alain Badiou claims that ontology is mathematics. Badiou also believes in deep connections between mathematics, poetry and philosophy.

In some cases, natural philosophers and other scientists who have made extensive use of mathematics have made leaps of inference between beauty and physical truth in ways that turned out to be erroneous. For example, at one stage in his life, Johannes Kepler believed that the proportions of the orbits of the then-known planets in the Solar System have been arranged by God to correspond to a concentric arrangement of the five Platonic solids, each orbit lying on the circumsphere of one polyhedron and the insphere of another. As there are exactly five Platonic solids, Kepler's hypothesis could only accommodate six planetary orbits and was disproved by the subsequent discovery of Uranus.

35.5   Beauty and mathematical information theory

In the 1970s, Abraham Moles and Frieder Nake analyzed links between beauty, information processing, and information theory.[14][15] In the 1990s, Jürgen Schmidhuber formulated a mathematical theory of observer-dependent subjective beauty based on algorithmic information theory: the most beautiful objects among subjectively comparable objects have short algorithmic descriptions (i.e., Kolmogorov complexity) relative to what the observer already knows.[16][17][18] Schmidhuber explicitly distinguishes between beautiful and interesting. The latter corresponds to the first derivative of subjectively perceived beauty: the observer continually tries to improve the predictability and compressibility of the observations by discovering regularities such as repetitions and symmetries and fractal self-similarity. Whenever the observer's learning process (possibly a predictive artificial neural network) leads to improved data compression such that the observation sequence can be described by fewer bits than before, the temporary interestingness of the data corresponds to the compression progress, and is proportional to the observer's internal curiosity reward[19][20]

35.6   Mathematics and the arts

Main articles: Mathematics and art and Mathematics and music

35.6.1   Music

Examples of the use of mathematics in music include the stochastic music of Iannis Xenakis, counterpoint of Johann Sebastian Bach, polyrhythmic structures (as in Igor Stravinsky's *The Rite of Spring*), the Metric modulation of Elliott Carter, permutation theory in serialism beginning with Arnold Schoenberg, and application of Shepard tones in Karlheinz Stockhausens *Hymnen*.

35.6.2   Visual arts

Examples of the use of mathematics in the visual arts include applications of chaos theory and fractal geometry to computer-generated art, symmetry studies of Leonardo da Vinci, projective geometries in development of the perspective theory of Renaissance art, grids in Op art, optical geometry in the camera obscura of Giambattista della Porta, and multiple perspective in analytic cubism and futurism.

The Dutch graphic designer M.C. Escher created mathematically inspired woodcuts, lithographs, and mezzotints. These feature impossible constructions, explorations of infinity, architecture, visual paradoxes and tessellations. British constructionist artist John Ernest created reliefs and paintings inspired by group theory.[21] A number of other British artists of the constructionist and systems schools also draw on mathematics models and structures as a source of inspiration, including Anthony Hill and Peter Lowe. Computer-generated art is based on mathematical algorithms.

35.7   See also

- Argument from beauty

- Cellular automaton

- Descriptive science

- Fluency heuristic

- Golden ratio

- Mathematics and architecture

- Normative science

- Philosophy of mathematics

- Processing fluency theory of aesthetic pleasure

- Pythagoreanism

- Theory of everything

35.8   Notes

[1] Russell, Bertrand (1919). "The Study of Mathematics". *Mysticism and Logic: And Other Essays*. Longman. p. 60. Retrieved 2008-08-22.

[2] Devlin, Keith (2000). "Do Mathematicians Have Different Brains?". *The Math Gene: How Mathematical Thinking Evolved And Why Numbers Are Like Gossip*. Basic Books. p. 140. ISBN 978-0-465-01619-8. Retrieved 2008-08-22.

[3] Elisha Scott Loomis published over 360 proofs in his book Pythagorean Proposition (ISBN 0-873-53036-5).

[4] Rota (1997), *The phenomenology of mathematical beauty*, p. 173

[5] Gallagher, James (13 February 2014). "Mathematics: Why the brain sees maths as beauty". *BBC News online*. Retrieved 13 February 2014.

[6] Feynman, Richard P. (1977). *The Feynman Lectures on Physics* I. Addison-Wesley. p. 22-10. ISBN 0-201-02010-6.

[7] Hardy, G.H. "18". Missing or empty |title= (help)

[8] Rota, *The phenomenology of mathematical beautyyear = 1997*, p. 172

[9] Monastyrsky (2001), *Some Trends in Modern Mathematics and the Fields Medal*

[10] Lang, p. 3

[11] Chandrasekhar, p. 148

[12] Phillips, George (2005). "Preface". *Mathematics Is Not a Spectator Sport*. Springer Science+Business Media. ISBN 0-387-25528-1. Retrieved 2008-08-22. "...there is nothing in the world of mathematics that corresponds to an audience in a concert hall, where the passive listen to the active. Happily, mathematicians are all *doers*, not spectators.

[13] Schechter, Bruce (2000). *My brain is open: The mathematical journeys of Paul Erdős*. New York: Simon & Schuster. pp. 70–71. ISBN 0-684-85980-7.

[14] A. Moles: *Théorie de l'information et perception esthétique*, Paris, Denoël, 1973 (Information Theory and aesthetical perception)

[15] F Nake (1974). Ästhetik als Informationsverarbeitung. (Aesthetics as information processing). Grundlagen und Anwendungen der Informatik im Bereich ästhetischer Produktion und Kritik. Springer, 1974, ISBN 3-211-81216-4, ISBN 978-3-211-81216-7

[16] J. Schmidhuber. Low-complexity art. Leonardo, Journal of the International Society for the Arts, Sciences, and Technology, 30(2):97–103, 1997. http://www.jstor.org/pss/1576418

[17] J. Schmidhuber. Papers on the theory of beauty and low-complexity art since 1994: http://www.idsia.ch/~{}juergen/beauty.html

[18] J. Schmidhuber. Simple Algorithmic Principles of Discovery, Subjective Beauty, Selective Attention, Curiosity & Creativity. Proc. 10th Intl. Conf. on Discovery Science (DS 2007) p. 26-38, LNAI 4755, Springer, 2007. Also in Proc. 18th Intl. Conf. on Algorithmic Learning Theory (ALT 2007) p. 32, LNAI 4754, Springer, 2007. Joint invited lecture for DS 2007 and ALT 2007, Sendai, Japan, 2007. http://arxiv.org/abs/0709.0674

[19] .J. Schmidhuber. Curious model-building control systems. International Joint Conference on Neural Networks, Singapore, vol 2, 1458–1463. IEEE press, 1991

[20] Schmidhuber's theory of beauty and curiosity in a German TV show: http://www.br-online.de/bayerisches-fernsehen/faszination-wissen/schoenheit--aesthetik-wahrnehmung-ID1212005092828.xml

[21] John Ernest's use of mathematics and especially group theory in his art works is analysed in *John Ernest, A Mathematical Artist* by Paul Ernest in Philosophy of Mathematics Education Journal, No. 24 Dec. 2009 (Special Issue on Mathematics and Art): http://people.exeter.ac.uk/PErnest/pome24/index.htm

35.9   References

- Aigner, Martin, and Ziegler, Gunter M. (2003), *Proofs from THE BOOK,* 3rd edition, Springer-Verlag.

- Chandrasekhar, Subrahmanyan (1987), *Truth and Beauty: Aesthetics and Motivations in Science,* University of Chicago Press, Chicago, IL.

- Hadamard, Jacques (1949), *The Psychology of Invention in the Mathematical Field,* 1st edition, Princeton University Press, Princeton, NJ. 2nd edition, 1949. Reprinted, Dover Publications, New York, NY, 1954.

- Hardy, G.H. (1940), *A Mathematician's Apology,* 1st published, 1940. Reprinted, C.P. Snow (foreword), 1967. Reprinted, Cambridge University Press, Cambridge, UK, 1992.

- Hoffman, Paul (1992), *The Man Who Loved Only Numbers,* Hyperion.

- Huntley, H.E. (1970), *The Divine Proportion: A Study in Mathematical Beauty,* Dover Publications, New York, NY.

- Loomis, Elisha Scott (1968), *The Pythagorean Proposition,* The National Council of Teachers of Mathematics. Contains 365 proofs of the Pythagorean Theorem.

- Lang, Serge (1985). *The Beauty of Doing Mathematics: Three Public Dialogues.* New York: Springer-Verlag. ISBN 0-387-96149-6.

- Peitgen, H.-O., and Richter, P.H. (1986), *The Beauty of Fractals,* Springer-Verlag.

- Reber, R., Brun, M., & Mitterndorfer, K. (2008). The use of heuristics in intuitive mathematical judgment. *Psychonomic Bulletin & Review, 15,* 1174-1178.

- Strohmeier, John, and Westbrook, Peter (1999), *Divine Harmony, The Life and Teachings of Pythagoras,* Berkeley Hills Books, Berkeley, CA.

- Rota, Gian-Carlo (1997). "The phenomenology of mathematical beauty". *Synthese* 111 (2): 171–182. doi:10.1023/A:100493072223

- Monastyrsky, Michael (2001). "Some Trends in Modern Mathematics and the Fields Medal" (PDF). *Can. Math. Soc. Notes* 33 (2 and 3).

35.10 External links

- Mathematics, Poetry and Beauty

- Is Mathematics Beautiful?

- The Beauty of Mathematics

- Justin Mullins

- Edna St. Vincent Millay (poet): *Euclid alone has looked on beauty bare*

- Terence Tao, *What is good mathematics?*

- Mathbeauty Blog

- The *Aesthetic Appeal* collection at the Internet Archive

- *A Mathematical Romance* Jim Holt December 5, 2013 issue of The New York Review of Books review of *Love and Math: The Heart of Hidden Reality* by Edward Frenkel

Chapter 36

Proofs from THE BOOK

Proofs from THE BOOK is a book of mathematical proofs by Martin Aigner and Günter M. Ziegler. The book is dedicated to the mathematician Paul Erdős, who often referred to "The Book" in which God keeps the most elegant proof of each mathematical theorem. During a lecture in 1985, Erdős said, "You don't have to believe in God, but you should believe in The Book."

Proofs from THE BOOK contains 32 sections (44 in the fifth edition), each devoted to one theorem but often containing multiple proofs and related results. It spans a broad range of mathematical fields: number theory, geometry, analysis, combinatorics and graph theory. Erdős himself made many suggestions for the book, but died before its publication. The book is illustrated by Karl Heinrich Hofmann. It has gone through five editions in English, and has been translated into Persian, French, German, Hungarian, Italian, Japanese, Chinese, Polish, Portuguese, Korean, Turkish, Russian and Spanish.

The proofs include:

- Proof of Bertrand's postulate

- Proof that e is irrational (also showing the irrationality of certain related numbers)

- Six proofs of the infinitude of the primes, including Euclid's and Furstenberg's

- Monsky's theorem (4th edition)

- Wetzel's problem on families of analytic functions with few distinct values

36.1   References

- *Proofs from the book*. Springer, Berlin 1998, ISBN 3-540-63698-6

  - Aigner, Martin; Ziegler, Günter (2009). *Proofs from THE BOOK* (4th ed.). Berlin, New York: Springer-Verlag. ISBN 978-3-642-00855-9.
  - Table of Contents of the 4th ed.

- Günter M. Ziegler's homepage, including a list of editions and translations.

- Shepard, Mary (1999). "Read This! Review of Proofs from THE BOOK". Mathematical Association of America..

36.2 Text and image sources, contributors, and licenses

36.2.1 Text

- **Mathematical proof** *Source:* https://en.wikipedia.org/wiki/Mathematical_proof?oldid=665905684 *Contributors:* AxelBoldt, Robert Merkel, The Anome, Toby Bartels, Youandme, Edward, TeunSpaans, Michael Hardy, Spazlink, TakuyaMurata, LittleDan, Pratyeka, Glenn, Tim Retout, Hashar, Revolver, Charles Matthews, Dysprosia, Jitse Niesen, Bjh21, Markhurd, Booya, Bcorr, Denelson83, Fredrik, Romanm, Lowellian, Gandalf61, Paul G, Hadal, Giftlite, Markus Krötzsch, Yekrats, CSTAR, Karl-Henner, Tyler McHenry, Hkpawn~enwiki, Petershank, Peter Kwok, Andreas Kaufmann, PhotoBox, Jason Carreiro, Leibniz, Paul August, Robertbowerman, Syp, El C, Skeppy, Stesmo, Jojit fb, Nk, ACW, Jumbuck, Kanie, Arthena, Itsmine, Oleg Alexandrov, Jslu, Mindmatrix, Rodrigo Rocha~enwiki, Bkkbrad, Pol098, Jeff3000, Tygar, Stevey7788, Seyon, Rjwilmsi, Salix alba, R.e.b., Alexb@cut-the-knot.com, KarlFrei, Jrtayloriv, BMF81, Chobot, Jersey Devil, DVdm, Borgx, Trovatore, Cheeser1, Boivie, FF2010, PTSE, Josh3580, HereToHelp, Willtron, Curpsbot-unicodify, Brentt, ⟨⟨⟨⟨ robot, SmackBot, KnowledgeOf-Self, Bomac, Jagged 85, Diegotorquemada, Gracenotes, Can't sleep, clown will eat me, Rantingsteve, Rrburke, Drackap, MichaelBillington, Occultations, Luke Gustafson, Tparameter, Policron, Jonjesbuzz, Dessources, Vanished user 39948282, Idioma-bot, VolkovBot, JohnBlackburne, Jimmaths, Greatwalk, TXiKiBoT, Gentlemath, Anonymous Dissident, Ocolon, AlleborgoBot, Symane, EmxBot, SieBot, YonaBot, Phe-bot, RiskAverse, RJaguar3, Triwbe, X-Fi6, LeadSongDog, 360 Degree, Barliner, Radon210, Byrialbot, Thehotelambush, DesolateReality, Nusumareta, Randomblue, Myrvin, Tautologist, Athenean, ClueBot, Justin W Smith, Dobermanji, Lozersk, Adrianwn, ChandlerMapBot, I am a violinist, Wikiimee, DragonBot, Alexbot, Alejandrocaro35, La Pianista, Humanengr, Pichpich, Stickee, Gerhardvalentin, Kwjbot, WikiDao, EEng, Addbot, Betterusername, Glane23, AndersBot, Roux, ChenzwBot, Jaydec, Numbo3-bot, BOOLE1847, Legobot, Luckas-bot, Yobot, Kan8eDie, Jorge-Fierro, Texhausballa, AnomieBOT, JRB-Europe, Materialscientist, Are you ready for IPv6?, La comadreja, ArthurBot, LilHelpa, MauritsBot, Xqbot, Sourceholder, Almabot, J04n, Geometryfan, FrescoBot, LucienBOT, Pinethicket, FriedrickMILBarbarossa, Foodimentary, Di1000, Jauhienij, Leasnam, FoxBot, TobeBot, DixonDBot, Robert hoffman, EmausBot, AmigoDoPaulo, Josve05a, Sven Manguard, Jijo925, ClueBot NG, Wikigold96, CocuBot, Melville88, Widr, راجح أحمد, Keetanii, Helpful Pixie Bot, Joolsa123, Wasbeer, Khonkhortisan, Kiewbra, Hebert Peró, Was123ification, Dexbot, Mogism, Brirush, Epicgenius, Gdaniel111, Jamesmcmahon0, Purnendu Karmakar, Schopenhauersville, Blackbombchu, Coz7, Seansmoove27, Leegrc, EggyEggPercent, TheCoffeeAddict and Anonymous: 189

- **Philosophy of mathematics** *Source:* https://en.wikipedia.org/wiki/Philosophy_of_mathematics?oldid=671116977 *Contributors:* Damian Yerrick, AxelBoldt, Matthew Woodcraft, Derek Ross, LC~enwiki, Bryan Derksen, Zundark, The Anome, Hhanke, SimonP, Zadcat, Ryguasu, Cwitty, IanS, The hanged man, Michael Hardy, Nixdorf, BoNoMoJo (old), Gabbe, Chinju, GTBacchus, Dori, Eric119, Ahoerstemeier, Snoyes, Darkwind, Cyan, Tim Retout, Rotem Dan, Andres, EdH, Schneelocke, Renamed user 4, Charles Matthews, RickK, Wu, Dtgm, Markhurd, Maximus Rex, Unknown, Robbot, Romanm, Gandalf61, Tim Ivorson, MathMartin, OmegaMan, Hadal, JohannesMarat, Aetheling, Tobias Bergmann, Adam78, Giftlite, Lee J Haywood, Lethe, Monedula, Everyking, Esap, WHEELER, Just Another Dan, JRR Trollkien, Stevietheman, Alberto da Calvairate~enwiki, Karol Langner, Rdsmith4, Pmanderson, Robin klein, Random account 47, Shahab, D6, Splatty, Rich Farmbrough, Wclark, YUL89YYZ, Paul August, Pban92, Elwikipedista~enwiki, Syp, Lycurgus, Rgdboer, Episcopo, Pokereth, Root4(one), Pearle, IonNerd, Jumbuck, Vesal, Droob, Eagleamn, Ossiemanners, Sligocki, Mr. Hyde~enwiki, Hu, Pernest2002, Bookandcoffee, Euphrosyne, Angr, Boothy443, Mel Etitis, Woohookitty, Linas, Barrylb, Jok2000, Al E., Dfranke, DaveApter, BD2412, Qwertyus, Gigapixel, Rjwilmsi, Salix alba, Mathbot, Nihiltres, Echeneida, RexNL, Mark J, David H Braun (1964), JamesLee, YurikBot, Wavelength, Shonk, Hairy Dude, Conscious, Hydrargyrum, Gaius Cornelius, Chaos, KSchutte, Aldux, Hakeem.gadi, Tomisti, Igiffin, Arthur Rubin, Skullfission, LeonardoRob0t, Canadianism, Meegs, Infinity0, Brentt, Sardanaphalus, JJL, SmackBot, Reedy, Tom Lougheed, Pokipsy76, Eskimbot, Gilliam, Ghosts&empties, Chris the speller, Bluebot, Trebor, JMSwtlk, Helder Ribeiro, Go for it!, Lesnail, Cybercobra, John wesley, Jon Awbrey, Just plain Bill, Bidabadi~enwiki, Vina-iwbot~enwiki, FlyHigh, Clicketyclack, Byelf2007, Igrant, Dbtfz, JHunterJ, Mets501, Ryulong, Iridescent, K. Aeternus, Rhetth, Tawkerbot2, 8754865, CRGreathouse, CmdrObot, CBM, Pan Camel, Gregbard, Logicus, Danman3459, Peterdjones, M a s, JamesAM, Old port, Thijs!bot, W3asal, 271828182, Headbomb, Marek69, Nick Number, Klausness, Vodello, Rjmars97, GeePriest, Wayiran, Leuqarte, Tayl1257, Opbion, Huphelmeyer, Andrewthomas10, Wlod, Lucaas, Revery~enwiki, Seberle, David Eppstein, Martynas Patasius, Nowletsgo, Ynotds, CommonsDelinker, Filll, Altes, Maurice Carbonaro, NerdyNSK, Dispenser, Mikael Häggström, Rnest2002, Infarom, Milogardner, AlnoktaBOT, Jimmaths, TXiKiBoT, Aleph42, The Tetrast, Philogo, Broadbot, Geometry guy, Popopp, Finalfantasy2012, Wenli, Tomaxer, Sapphic, Dmcq, Newbyguesses, SieBot, Darrell Wheeler, Lightmouse, Roran659, DesolateReality, Classicalecon, ClueBot, DFRussia, Rockfang, CohesionBot, Azadeh.a, Sun Creator, Vegetator, JKeck, Marc van Leeuwen, Gerhardvalentin, Zodon, Addbot, Fyrael, With goodness in mind, MrOllie, Imtg5102, ProfessorThunderlips, Drdonzi, Nallimbot, Synchronism, AnomieBOT, Materialscientist, Citation bot, Xqbot, Capricorn42, Crzer07, Uarrin, J04n, Freddyfirre, Omnipaedista, Taekwandean, Aaron Kauppi, Constructive editor, Mr fabs, Hugetim, FrescoBot, Mark Renier, EricAndrewWallace, BrideOfKripkenstein, Alboran, Steve Quinn, CESSMASTER, Machine Elf 1735, Mary rose arias, Tkuvho, Pinethicket, Pollinosisss, Jdapayne, Vrenator, Hueyha, Jowa fan, Merehap, EmausBot, John of Reading, ZéroBot, Amacfiew, RaptureBot, HarmoniousMembrane, BartlebytheScrivener, RockMagnetist, Logicalgregory, Anita5192, E. Fokker, ClueBot NG, Alexander E Ross, Deer*lake, Helpful Pixie Bot, Beaumont877, Brian Tomasik, Harizotoh9, Brad7777, Gibbja, Dexbot, Jochen Burghardt, Mark viking, Stara729, Workstern, Kyle1009, Waluigigod, Cnbr15 and Anonymous: 225

- **Language of mathematics** *Source:* https://en.wikipedia.org/wiki/Language_of_mathematics?oldid=662393256 *Contributors:* Bdesham, Snoyes, Nikai, Charles Matthews, Bevo, Gandalf61, Giftlite, Waltpohl, Marcos, Andreas Kaufmann, Leibniz, Xezbeth, Gauge, CanisRufus, Rgdboer, I9Q79oL78KiL0QTFHgcy, Mdd, Diego Moya, Oleg Alexandrov, Woohookitty, Splintax, SixWingedSeraph, Kbdank71, Bgwhite, Katieh5584, SmackBot, Bluebot, A Geek Tragedy, Rrburke, Lambiam, CRGreathouse, Gregbard, Sluzzelin, Cic, David Eppstein, SuneJ~enwiki, Desolate-Reality, Foxj, Rjd0060, CohesionBot, Muhandes, Dmorsund, Addbot, Jarble, AnomieBOT, Xqbot, Jebdm, Kierkkadon, TRBP, GoingBatty, Fixblor, Bluefairyturnedred, PhantomTech, Gst.steven and Anonymous: 29

- **Formal language** *Source:* https://en.wikipedia.org/wiki/Formal_language?oldid=662891308 *Contributors:* LC~enwiki, Jan Hidders, Andre Engels, Youandme, Rp, Ahoerstemeier, Nanshu, Schneelocke, Charles Matthews, Bernoeial, Hyacinth, Cleduc, Spikey, Robbot, Jaredwf,

Millosh, Tobias Bergemann, Connelly, Giftlite, Kim Bruning, Eequor, ConradPino, MarkSweep, Mukerjee, Tyler McHenry, Discospinster, Rich Farmbrough, Dbachmann, Ntennis, Chalst, Nk, Jonsafari, Obradovic Goran, Waku, Helix84, Mdd, Jumbuck, MrTree, Stephan Leeds, Nuno Tavares, Linas, Ruud Koot, Flamingspinach, ThomasOwens, Sdornan, LKenzo, FlaBot, Eubot, PlatypeanArchcow, Pexatus, Chobot, MithrandirMage, YurikBot, Wavelength, RussBot, Hede2000, Archelon, Rick Norwood, Muu-karhu, Bkil, Aaron Schulz, BOTSuperzerocool, Bota47, Saric, Ripper234, Arthur Rubin, Jogers, GrinBot~enwiki, TuukkaH, Finell, SmackBot, Wic2020, Incnis Mrsi, Unyoyega, Od Mishehu, Fikus, Jpvinall, Jerome Charles Potts, J. Spencer, Frap, Cerebralpayne, Jon Awbrey, FelisLeo, SashatoBot, Astuishin, Mike Fikes, Iridescent, Dreftymac, Adriatikus, DBooth, CRGreathouse, CBM, Ezrakilty, Gregbard, Alaibot, NERIUM, Nick Number, Cyclonenim, Vantelimus, Danny lost, VictorAnyakin, Hermel, AndriesVanRenssen, Nyq, JNW, Tedickey, Cic, David Eppstein, J.delanoy, Trusilver, Brest, Daniele.tampieri, Policron, Bonadea, Halukakin, Idioma-bot, Philomathoholic, VolkovBot, AlnoktaBOT, Philogo, ASHPvanRenssen, AlleborgoBot, Newbyguesses, MiNombreDeGuerra, Fratrep, OKBot, CBM2, Classicalecon, ClueBot, Alpha Beta Epsilon, Excirial, Quercus basaseachicensis, Alexbot, NuclearWarfare, Hans Adler, Addbot, Cuaxdon, LemmeyBOT, OlEnglish, Zorrobot, Jarble, JakobVoss, Legobot, Luckas-bot, Yobot, Pcap, AnomieBOT, Jim1138, JackieBot, Hahahaha4, Citation bot, Clickey, Xqbot, GrouchoBot, Charvest, Wei.cs, Serberimor, MastiBot, RobinK, Full-date unlinking bot, ActivExpression, Gzorg, Lokentaren, LoStrangolatore, Mean as custard, Ankog, Dziadgba, Architectchao, Tijfo098, ClueBot NG, Helpful Pixie Bot, Garsd, Sff9, ChrisGualtieri, Dmunene, Chunliang Lyu, Dulaambaw, Akerbos, Jochen Burghardt, Isthatmoe, KasparBot and Anonymous: 87

- **Elementary proof** *Source:* https://en.wikipedia.org/wiki/Elementary_proof?oldid=474298901 *Contributors:* Michael Hardy, Charles Matthews, Sam nead, Paul August, Alansohn, Greg Kuperberg, Oleg Alexandrov, BD2412, R.e.b., KarlFrei, Star trooper man, Finell, Timotheus Canens, Kukini, Lambiam, JoshuaZ, Braddodson, Myasuda, Doctormatt, Cydebot, Ntsimp, Headbomb, Dalassa, David Eppstein, Atheuz, ClueBot, Citation bot 1 and Anonymous: 2

- **Formal proof** *Source:* https://en.wikipedia.org/wiki/Formal_proof?oldid=661169170 *Contributors:* Charles Matthews, Hyacinth, J D, Timrollpickering, Pmanderson, EmilJ, Dfranke, BD2412, Salix alba, Gaius Cornelius, Seegoon, Chrylis, Gregbard, Nick Number, Philogo, VanishedUserABC, Radagast3, Ndenison, Mild Bill Hiccup, Hans Adler, HexaChord, Addbot, Numbo3-bot, AnomieBOT, MattTait, The Wiki ghost, Disinvented and Anonymous: 9

- **Proof theory** *Source:* https://en.wikipedia.org/wiki/Proof_theory?oldid=630512528 *Contributors:* Mav, Bryan Derksen, The Anome, Toby Bartels, Youandme, Michael Hardy, Llywrch, Dominus, Rotem Dan, Charles Matthews, Dysprosia, Markhurd, Hyacinth, Jni, Giftlite, Markus Krötzsch, Jorend, Kntg, Leibniz, Pj.de.bruin, Luqui, Number 0, Brian0918, Chalst, Nortexoid, Msh210, Krappie, Wtmitchell, Gene Nygaard, Oleg Alexandrov, Jtauber, Porcher, Mathbot, Comiscuous, Tillmo, Chobot, YurikBot, Hairy Dude, Tong~enwiki, Arthur Rubin, Sardanaphalus, SmackBot, Cabe6403, Jahiegel, Vina-iwbot~enwiki, Byelf2007, Lambiam, Dbtfz, Kuru, Rizome~enwiki, Dicklyon, JRSpriggs, CBM, Gregbard, Nick Number, Escarbot, LaForge, David Eppstein, Yonaa, J.delanoy, Policron, Alan U. Kennington, JohnBlackburne, Hqb, Qxz, Magmi, VanishedUserABC, Radagast3, OKBot, Kumioko, Anchor Link Bot, CBM2, ClueBot, Mannypabla, Thisthat12345, Beach drifter, Good Olfactory, Addbot, Matěj Grabovský, Dima125, Yobot, Ptbotgourou, JRB-Europe, MattTait, Citation bot, Tbvdm, Sa'y, Foobarnix, Gamewizard71, Dbmikus, WildBot, EmausBot, ZéroBot, Donner60, ClueBot NG, Rezabot, Helpful Pixie Bot, Brad7777, Rsmbf, Begadkepat, Kkval0, JHU1959 and Anonymous: 46

- **Theorem** *Source:* https://en.wikipedia.org/wiki/Theorem?oldid=668391294 *Contributors:* AxelBoldt, Mav, Zundark, The Anome, Tarquin, Tbackstr, XJaM, Aldie, Michael Hardy, Zeno Gantner, TakuyaMurata, Bagpuss, Glenn, Tim Retout, Rotem Dan, Andres, Charles Matthews, Dcoetzee, Bemoeial, Hyacinth, Traroth, SirPeebles, Moriel~enwiki, Josh Cherry, Fredrik, MathMartin, Ojigiri~enwiki, Timrollpickering, Hadal, Alan Liefting, Snobot, Ancheta Wis, Tosha, Giftlite, Monedula, Frupuff, Fishal, Chowbok, Alaz, MarkSweep, Karol Langner, Jacob grace, Pmanderson, Tyler McHenry, Hkpawn~enwiki, Tzanko Matev, Joyous!, EugeneZelenko, Discospinster, Rich Farmbrough, Paul August, Bender235, Gauge, Tompw, El C, Edwinstearns, Billymac00, Sasquatch, Alansohn, Gary, Sciurinæ, Joriki, Igny, Ruud Koot, Erasmus, Mekong Bluesman, Tslocum, Graham87, BD2412, Gmelli, Sdornan, Salix alba, Jrtayloriv, AndriuZ, M7bot, Chobot, MithrandirMage, Sbrools, DVdm, Algebraist, Roboto de Ajvol, Borgx, Chaos, Trovatore, Dbfirs, Bota47, Tomisti, Arthur Rubin, Kier07, Anclation~enwiki, Curpsbot-unicodify, Erudy, Finell, Sardanaphalus, SmackBot, RDBury, Bomac, Skizzik, Chris the speller, Fuzzform, MalafayaBot, DHNbot~enwiki, Sholto Maud, Cybercobra, Acdx, SashatoBot, Lambiam, IronGargoyle, Craigblock, Lalaith, Autonova, Mike Fikes, Zero sharp, JRSpriggs, CRGreathouse, Hi.ro, CBM, Gregbard, Cydebot, R Harris, Moxmalin, Thijs!bot, Epbr123, Mrcs, James086, Nick Number, AntiVandalBot, Thenub314, .anacondabot, Magioladitis, Dvptl, Animum, Gwern, Stephenchou0722, Pomte, Hippasus the Younger, Fcsuper, Jeepday, Nznancy, Coppertwig, Haseldon, Tparameter, Fjbfour, Dessources, DavidCBryant, Caiodnh, Austinmohr, VolkovBot, Psmythirl, Am Fiosaigear~enwiki, TXiKiBoT, Rei-bot, IKiddo, Voorlandt, Philogo, Geometry guy, Graymornings, Dmcq, HiDrNick, Destroyerof-Dreams, Hthoreau2, Newbyguesses, SieBot, Respir, Huzzah018, Oxymoron83, SimonTrew, OKBot, Kumioko (renamed), DesolateReality, Wjemather, Loren.wilton, ClueBot, Blanchardb, Excirial, Alexbot, TheSnacks, Hans Adler, Muzz 2008, MonoBot, XLinkBot, Burkaja, Marc van Leeuwen, SilvonenBot, Badgernet, Addbot, DrThunder88, Some jerk on the Internet, CanadianLinuxUser, CarsracBot, Dr. Universe, Favonian, Numbo3-bot, Flatfish89, Stepfordswife, Zorrobot, Legobot, Cote d'Azur, Luckas-bot, Yobot, TaBot, II MusLiM HyBRiD II, Kristen Eriksen, LilHelpa, Xqbot, Doezxcty, Capricorn42, Almabot, Miym, GrouchoBot, Jubb-green, RibotBOT, FrescoBot, Sirtywell, Haeinous, Heptadecagon, BigDwiki, RedBot, Eric wisniewski, EmausBot, ZéroBot, The Nut, Xzenu, GZ-Bot, H3llBot, D.Lazard, ChuispastonBot, ClueBot NG, Helpful Pixie Bot, Howald, Jibun, bukiyou desu kara, Khazar2, Oracions, Wywin, Yamaha5, Loraof, EoRdE6, BabyChastie, Nbro, KasparBot, Blakktaktiks and Anonymous: 111

- **Axiom** *Source:* https://en.wikipedia.org/wiki/Axiom?oldid=671080640 *Contributors:* AxelBoldt, LC~enwiki, Brion VIBBER, Mav, Zundark, The Anome, Youssefsan, XJaM, Stevertigo, Michael Hardy, JakeVortex, Nixdorf, Graue, Glenn, Tim Retout, Nikai, Rotem Dan, Andres, Hectorthebat, EdH, Rob Hooft, Mxn, Charles Matthews, Dysprosia, Greenrd, Markhurd, Hyacinth, CecilTyme, Robbot, Josh Cherry, Red-Wolf, Altenmann, Jeronim, Flauto Dolce, Saforrest, Wikibot, Benc, Tobias Bergemann, Ancheta Wis, Tosha, Giftlite, Mshonle~enwiki, Gene Ward Smith, Dissident, Alterego, Neilc, Andycjp, Knutux, Alberto da Calvairate~enwiki, Antandrus, Dunks58, Sam Hocevar, Guppyfinsoup, Mormegil, EugeneZelenko, Rich Farmbrough, Iainscott, Mani1, Paul August, El C, Rgdboer, Irrᵗⁱᵒnal, Bobo192, Babomb, Mintywalker, ParticleMan, Nk, Rje, Obradovic Goran, Haham hanuka, Jumbuck, Alansohn, 119, Jeltz, Hu, Simplemelon, Stephan Leeds, HenryLi, Killing Vector, Oleg Alexandrov, Natalya, Roland2~enwiki, Kzollman, Ruud Koot, Gimboid13, Xeonx, Mandarax, Qwertyus, FreplySpang, Salix alba, Nmegill, FlaBot, Matharvest, Mathbot, Alvin-cs, Chobot, YurikBot, Wavelength, Splash, Chaos, Trovatore, Dbmag9, VinnyCee, Bucketsofg, Arthur Rubin, Vicarious, David Biddulph, Nahaj, SmackBot, McGeddon, Od Mishehu, SaxTeacher, Bomac, KocjoBot~enwiki, Alksub, RobotJcb, Bluebot, Adam M. Gadomski, DMS, MalafayaBot, SchfiftyThree, Spellchecker, Zhuravskij, Xiner, Xyzzyplugh, EPM, Pissant, Eddiesegoura, Jon Awbrey, Sammy1339, Vina-iwbot~enwiki, Byelf2007, SashatoBot, Lambiam, DA3N, Dialectic~enwiki, Physis, Atoll, 16@r,

Loadmaster, MTSbot~enwiki, Hu12, Dreftymac, Lenoxus, Bruno321, Happy-melon, Bharatveer, JRSpriggs, CRGreathouse, CaveBat, CBM, RoddyYoung, Gregbard, Cydebot, Peterdjones, Julian Mendez, Gproud, PamD, Fourchette, Ken Burns, CharlotteWebb, X96lee15, AlefZet, Escarbot, Quintote, Readro, Steelpillow, JAnDbot, Oxinabox, MER-C, Four Dog Night, Magioladitis, VoABot II, JamesBWatson, Swpb, Amorette, Daarznieks, David Eppstein, R'n'B, Trusilver, MistyMorn, Maurice Carbonaro, Amritasen7, Cpiral, SlowJog, McSly, Daniel5Ko, Juliancolton, Dorftrottel, VolkovBot, Hotfeba, Jeff G., LokiClock, Philip Trueman, TXiKiBoT, AlexDenney, Dmcq, Submachinegum, Sf-mammamia, LegoAxiom1007, SieBot, Dawn Bard, JerrySteal, MiNombreDeGuerra, IdNotFound, BrianGo28, Denisarona, DEMcAdams, ClueBot, Uncle Milty, Watchduck, Jotterbot, Eloifigueiredo, Aitias, Blackhawkjxx7, GAT27, YouRang?, JKeck, Marc van Leeuwen, Gnowor, Seshadri105, Addbot, Aelisabeths, NjardarBot, Tassedethe, Numbo3-bot, Dayewalker, Tide rolls, Legobot, Luckas-bot, Yobot, Fraggle81, TaBOT-zerem, KamikazeBot, Goldhawk08, Jalal0, AnomieBOT, Arjun G. Menon, Jim1138, Piano non troppo, LlywelynII, Bob Burkhardt, ArthurBot, Xqbot, LordDelta, J JMesserly, GrouchoBot, The Wiki ghost, Die4cheese5, Chjoaygame, FrescoBot, Paine Ellsworth, I dream of horses, RedBot, Xiglofre, Meier99, Gamewizard71, کاشف عقیل, Lotje, Dinamik-bot, 777sms, Duoduoduo, Reaper Eternal, Mini-mac, EmausBot, Najeeb1010, GoingBatty, Wikipelli, AvicBot, Jay-Sebastos, MonoAV, NTox, ClueBot NG, Masssly, Helpful Pixie Bot, Calabe1992, Frog23, Davidiad, Glacialfox, HumanNaturOriginal, Jeremy112233, ChrisGualtieri, Br'er Rabbit, Spectral sequence, Josell2, Lakshya08, Yolow1, Gcc333, Crystallizedcarbon, Cthulhu is love cthulhu is life, Tushar15mehta, Person420 and Anonymous: 271

- **Conjecture** *Source:* https://en.wikipedia.org/wiki/Conjecture?oldid=667618757 *Contributors:* AxelBoldt, The Anome, Fubar Obfusco, Stevertigo, Chas zzz brown, Michael Hardy, Ixfd64, Eric119, Ahoerstemeier, Glenn, Revolver, Charles Matthews, Timwi, Jm34harvey, Dysprosia, Gandalf61, Henrygb, Rasmus Faber, Mattflaschen, Tobias Bergemann, Ancheta Wis, Centrx, Giftlite, Andries, BenFrantzDale, Wikipedia benefits, Icairns, Hkpawn~enwiki, Yuriz, Chris Howard, Discospinster, Rich Farmbrough, Guanabot, 1pezguy, DcoetzeeBot~enwiki, El-wikipedista~enwiki, Porton, Bobo192, Obradovic Goran, Haham hanuka, Anthony Appleyard, Sligocki, Axeman89, Oleg Alexandrov, Grouchy-Dan, Justinlebar, Oliphaunt, Davidfstr, Graham87, BD2412, Qwertyus, Yurik, Eyu100, Salix alba, Tp, FlaBot, Mathbot, Kri, Gwernol, Algebraist, Wavelength, Sceptre, Ytcracker, Zwobot, Cadillac, GrinBot~enwiki, SmackBot, Jab843, Ohnoitsjamie, Ppntori, Bluebot, Hgrosser, BesselDekker, Byelf2007, Lambiam, Mouse Nightshirt, Valfontis, Scetoaux, Noah Salzman, Mr Stephen, George100, JForget, CRGreathouse, CBM, Gregbard, Doctormatt, Wang ty87916, Danny lost, JAnDbot, Andonic, Magioladitis, VoABot II, Careless hx, Juansidious, J.delanoy, McSly, Ohms law, Cometstyles, VolkovBot, Am Fiosaigear~enwiki, Hqb, DieBuche, Lova Falk, SieBot, Pallab1234, This, that and the other, ☺ ~enwiki, Antonio Lopez, Sjn28, Sunrise, Catrope, Escape Orbit, DEMcAdams, Mr. Granger, ClueBot, Mild Bill Hiccup, Blanchardb, Psychonomics, Cuckowski, NuclearWarfare, Aitias, Versus22, Qwfp, Editor2020, Cogitus1, Favonian, Qwrk, Legobot, Luckas-bot, Yobot, Fraggle81, THEN WHO WAS PHONE?, Reindra, AnomieBOT, AdjustShift, Phlembowper99, Larseven, Лев Дубовой, David Schwein, Lotje, EmausBot, Set theorist, ZéroBot, Stbunco, Trititaty, Rmashhadi, Garfieldperfect, ClueBot NG, Wcherowi, SusikMkr, LJosil, Startire, MerlIwBot, BlueMoonset, BG19bot, Johny Five, ChrisGualtieri, YFdyh-bot, Brirush, Purnendu Karmakar, MopSeeker, Jackmcbarn, AddWittyNameHere, Plesantdreams, Tcc astronaut, Loraof, Macosojr, Gavnerfsk and Anonymous: 144

- **Logic** *Source:* https://en.wikipedia.org/wiki/Logic?oldid=670139716 *Contributors:* AxelBoldt, Vicki Rosenzweig, The Anome, Toby Bartels, Ryguasu, Hirzel, Dwheeler, Stevertigo, Edward, Patrick, Chas zzz brown, Michael Hardy, Lexor, TakuyaMurata, Bagpuss, Looxix~enwiki, Ahoerstemeier, Notheruser, BigFatBuddha, Александър, Glenn, Marco Krohn, Rossami, Tim Retout, Rotem Dan, Evercat, EdH, Desert-Steve, Caffelice~enwiki, Mxn, Michael Voytinsky, Renamed user 4, Rzach, Charles Matthews, Dcoetzee, Paul Stansifer, Dysprosia, Jitse Niesen, Xiaodai~enwiki, Markhurd, MikeS, C Fenijn, SEWilco, Samsara, J D, Shizhao, Olathe, Jusjih, Ldo, Banno, Chuunen Baka, Robbot, Iwpg, Fredrik, R3m0t, Altenmann, MathMartin, Rorro, Rholton, Saforrest, Borislav, Robertoalencar, Michael Snow, Raeky, Guy Peters, Jooler, Filemon, Ancheta Wis, Exploding Boy, Giftlite, Recentchanges, Inter, Wolfkeeper, Lee J Haywood, COMPATT, Everyking, Rookkey, Malyctenar, Andris, Bovlb, Jason Quinn, Sundar, Siroxo, Deus Ex, Rheun, LiDaobing, Roachgod, Quadell, Starbane, Piotrus, Ludimer~enwiki, Karol Langner, CSTAR, Rdsmith4, APH, JimWae, OwenBlacker, Kntg, Mysidia, Pmanderson, Eduardoporcher, Eliazar, Grunt, Guppyfinsoup, Mike Rosoft, Freakofnurture, Ultratomio, Lorenzo Martelli, Discospinster, KillerChihuahua, Rhobite, Guanabot, Leibniz, Hippojazz, Vsmith, Raistlinjones, Slipstream, ChadMiller, Paul August, Bender235, El C, Chalst, Mwanner, Tverbeek, Bobo192, Cretog8, Johnkarp, Shenme, Amerindianarts, Passw0rd, Knucmo2, Storm Rider, Red Winged Duck, Alansohn, Shadikka, Rh~enwiki, Chira, ABCD, Kurt Shaped Box, SlimVirgin, Batmanand, Yummifruitbat, Shinjiman, Viella, Sciurinæ, MIT Trekkie, Alai, CranialNerves, Velho, Mel Etitis, Mindmatrix, Camw, Kokoriko, Kzollman, Ruud Koot, Orz, MONGO, Apokrif, Jok2000, Wikiklrsc, CharlesC, MarcoTolo, DRHansen, Gerbrant, Tslocum, Graham87, Alienus, BD2412, Porcher, Rjwilmsi, Mayumashu, Саша Стефановић, GOD, Bruce1ee, Salix alba, Crazynas, Ligulem, Baryonic Being, Titoxd, FlaBot, Kwhittingham, Latka, Mathbot, Twipley, SportsMaster, RexNL, AndriuZ, Quuxplusone, Celendin, Influence, R Lee E, JegaPRIME, Malhonen, Spencerk, Chobot, DVdm, EamonnPKeane, Roboto de Ajvol, Wavelength, Deeptrivia, KSmrq, Endgame~enwiki, Polyvios, CambridgeBayWeather, KSchutte, NawlinWiki, Rick Norwood, SEWilcoBot, Mipadi, Brimstone~enwiki, LaszloWalrus, AJHalliwell, Trovatore, Pontifexmaximus, Chunky Rice, Cleared as filed, Nick, Darkfred, Wjwma, Googl, Mendicott, StuRat, Open2universe, ChrisGriswold, Nikkimaria, OEMCUST, Nahaj, Extreme Unction, Allens, Sardanaphalus, Johndc, SmackBot, Lestrade, InverseHypercube, Pschelden, Jim62sch, Jagged 85, WookieInHeat, Josephprymak, Timotheus Canens, Srnec, UncreativeDrifter, Collingsworth, Gilliam, Skizzik, RichardClarke, Heliostellar, Chris the speller, Jaymay, Da nuke, Unbreakable MJ, MK8, Andrew Parodi, Kevin Hanse, MalafayaBot, Clconway, Sciyoshi~enwiki, Go for it!, Mikker, Zsinj, Can't sleep, clown will eat me, Misgnomer, Grover cleveland, Fuhghettaboutit, Cybercobra, Nakon, Jiddisch~enwiki, Richard001, MEJ119, Kabain52, Lacatosias, Jon Awbrey, DMacks, Henning Makholm, Ged UK, Cecil, Byelf2007, SashatoBot, Lambiam, Dbtfz, Deaconse, UberCryxic, FrozenMan, Heimstern, Shlomke, Shadowlynk, F15 sanitizing eagle, Prince153, WithstyleCMC, Hvn0413, Meco, RichardF, Novangelis, Vagary, Pamplmoose, KJS77, Hu12, Levineps, BananaFiend, K, Lottamiata, Catherineyronwode, Mrdthree, Igoldste, Themanofnines, Adambiswanger1, Satarnion, Tawkerbot2, Galex, SkyWalker, CRGreathouse, CBM, Editorius, Rubberchix, Gregbard, Kpossin, Cydebot, Pce3@ij.net, Jasperdoomen, Samuell, Quinnculver, Peterdjones, Travelbird, Pv2b, Drksl, JamesLucas, Julian Mendez, Dancter, Tawkerbot4, Shirulashem, Doug Weller, DumbBOT, Garik, Proginet, Mattisse, Letranova, Thijs!bot, Epbr123, Kredal, Smee, Marek69, AgentPeppermint, OrenBochman, Escarbot, Eleuther, Mentifisto, Vafthrudnir, AntiVandalBot, Peoppenheimer, Majorly, Gioto, Hidayat ullah, GeePriest, Dougher, Sluzzelin, JAnDbot, Narssarssuaq, MER-C, The Transhumanist, Avaya1, Zizon, Frankie816, Savant13, Dr mindbender, JJ119, JNW, Ann Matthias, Appraiser, Gamkiller, Smihael, Caesarjbsquitti, Midgrid, Bubba hotep, Moopiefoof, GeorgeFThomson, Virtlink, David Eppstein, Epsilon0, DerHexer, Waninge, Exbuzz, MartinBot, Wylve, CommonsDelinker, EdBever, C.R.Selvakumar, J.delanoy, Trusilver, Jbessie, Fictionpuss, Cpiral, RJ-Malko, McSly, Lightest~enwiki, Classicalsubjects, Mrg3105, Daniel5Ko, The Transhumanist (AWB), Policron, MetsFan76, Kenneth M Burke, Steel1943, Idioma-bot, Spellcast, WraithM, VolkovBot, Cireshoe, Rucha58, Macedonian, Hotfeba, Indubitably, Fundamental metric tensor, Jimmaths, Djhmoore, Aesopos, Rei-bot, Llamabr, Ontoraul, Philogo, Leafyplant, Sanfranman59, Abdullais4u, Cullowheean, Wikiisawesome,

Maxim, Myscience, LIBLAHLIBLAHTIMMAH, Synthebot, Rurik3, Koolo, Nagy, Symane, PGWG, W4chris, Prom2008, NHRHS2010, Radagast3, Demmy, JonnyJD, Newbyguesses, Linguist1, SieBot, StAnselm, Maurauth, Gerakibot, RJaguar3, Yintan, Bjrslogii, Soler97, Til Eulenspiegel, Flyer22, DanEdmonds, Undead Herle King, Crowstar, Redmarkviolinist, Spinethetic, Thelogicthinker, DancingPhilosopher, Svick, Valeria.depaiva, Adhawk, Sginc, Tognopop, CBM2, 3rdAlcove, PsyberS, Francvs, Classicalecon, Athenean, Atif.t2, Martarius, Clue-Bot, Andrew Nutter, Snigbrook, The Thing That Should Not Be, Taroaldo, Ukabia, TheOldJacobite, Boing! said Zebedee, Niceguyedc, Blanchardb, DragonBot, Jessieslame, Excirial, Alexbot, Jusdafax, Watchduck, AENAON, NuclearWarfare, Arjayay, SchreiberBike, Thingg, JDPhD, Scalhotrod, Budelberger, Skunkboy74, Gerhardvalentin, Duncan, Saeed.Veradi, Mcgauley08, NellieBly, Noctibus, Aunt Entropy, Jj-fuller123, Spidz, Addbot, Rdanneskjold, Proofreader77, Atethnekos, Sully111, Logicist, Vitruvius3, Glane23, Uber WoMensch!, Chzz, Favo-nian, LinkFA-Bot, AgadaUrbanit, Numbo3-bot, Ehrenkater, Tide rolls, Lightbot, Macro Shell, Zorrobot, Jarble, JEN9841, Aarsalankhalid, GorgeUbuasha, Arcvirgos 08, Jammie101, Francos22, Azcolvin429, MassimoAr, AnomieBOT, Jim1138, IRP, AdjustShift, Melune, NickK, Neurolysis, ArthurBot, Gemtpm, Blueeyedbombshell, Junho7391, Xqbot, Gilo1969, The Land Surveyor, Tyrol5, A157247, Petropoxy (Lith-oderm Proxy), Uarrin, GrouchoBot, Hifcelik, Omnipaedista, פריימן,26ى78.פ‏ريمن‎, Tales23, GhalyBot, Aaron Kauppi, GliderMaven, FrescoBot, Liridon, D'Artagnol, Tobby72, D'ohBot, Mewulwe, Itisnotme, Rhalah, Citation bot 1, Chenopodiaceous, AstaBOTh15, Gus the mouse, Pinethicket, Vicenarian, A8UDI, Ninjasaves, Seryred123, Maokart444, Gamewizard71, FoxBot, TobeBot, Burritoburritoburrito, Mys-ticcooperfox, Lotje, GregKaye, Vistascan, Vrenator, Duoduoduo, גרשון בן, Merlinsorca, Literateur, Jarpup, Whisky drinker, Mean as custard, Rlnewma, TjBot, Walkinxyz, EmausBot, Orphan Wiki, The Kytan Apprentice, Flosfa, Chrisct1993, Brad7777, Mewhho18, A.coolmcfly, Compulogger, Cyberbot II, Roger Smalling, The Illusive Man, NanishaOpaenyak, Rhlozier, EagerToddler39, Dexbot, Marius siuram, Табалдыев Ысламбек, Omanchandy007, RideLightning, Jochen Burghardt, Wieldthespade, Hippocamp, Wickid123, Matticusmadness, JMCF125, NIXONDIXON, CsDix, I am One of Many, ללעד נייר, Tentinator, EvergreenFir, Babitaarora, Ugog Nizdast, Melody Lavender, JustBerry, Skansi.sandro, Ginsuloft, Robf00f1235, Calvinator8, The Annoyed Logician, Liz, GreyWinterOwl, ByDash, Jbob13, Henniepenny, Matthew Derick B Cruz, Filedelinkerbot, Sher-lock502, Fvdedphill, Norwo037, Karnaoui, Pat132, The Expedia, Sbcdave, Muneeb Masoud, Jacksplay, Asdklf;, Esicam, Ntuser123, Cthulhu is love cthulhu is life, Adamrobson28, Josmust222, Layfi, KcBessy, SamiLayfi, Lanzdsey, SoSivr, Human3015, ZanderEdmunds, KasparBot, Bestusername-ign, Sparky Macgillicuddy, Mithisharma and Anonymous: 756

- **Informal logic** Source: https://en.wikipedia.org/wiki/Informal_logic?oldid=640996111 Contributors: Chris Q, DennisDaniels, Kwertii, Pcb21, Docu, Julesd, Dysprosia, Markhurd, Furrykef, Hyacinth, Tomchiukc, Leonard G., CSTAR, Paulscrawl, Guppyfinsoup, Chalst, Fernto, A.t.bruland, Velho, Simetrical, Mindmatrix, Rjwilmsi, Gurch, Mattpeck, Tsch81, NawlinWiki, NostinAdrek, Closedmouth, GraemeL, SmackBot, Stev0, Bluebot, Colonies Chris, Jon Awbrey, Byelf2007, Grumpyyoungman01, Meco, JohnsonRalph, Moreschi, Gregbard, Kpossin, Steel, Al Lemos, Escarbot, Doremítzwr, WinBot, Deeplogic, Clan-destine, Cathalwoods, Gomm, Ewen, R'n'B, RickardV, Adedayoojo, Homo logos, Nburden, Igglebop, Guillaume2303, Ontoraul, GlassFET, Kumioko (renamed), ClueBot, Napzilla, Staticshakedown, Addbot, SpBot, Lightbot, Luckas-bot, Luce nordica, AnomieBOT, LilHelpa, Omnipaedista, Gerald Roark, FrescoBot, LucienBOT, Machine Elf 1735, Abductive, Wotnow, John of Reading, ZéroBot, Tijfo098, ClueBot NG, Helpful Pixie Bot, GabeIglesia and Anonymous: 38

- **Deductive reasoning** Source: https://en.wikipedia.org/wiki/Deductive_reasoning?oldid=668159391 Contributors: The Cunctator, Toby Bar-tels, Youandme, Mrwojo, DennisDaniels, Michael Hardy, TakuyaMurata, BenKovitz, EdH, DesertSteve, Charles Matthews, Dtgm, Hyacinth, Lumos3, Robbot, R3m0t, Romanm, AceMyth, Blainster, Tobias Bergemann, Ancheta Wis, Giftlite, Lethe, Guanaco, Bovlb, Archenzo, Ja-son Quinn, Piotrus, Karol Langner, Aequo, Stepp-Wulf, EricBright, TedPavlic, Kevinb, Stbalbach, PhilHibbs, Causa sui, Flammifer, Jum-buck, Ryanmcdaniel, CyberSkull, Nasukaren, Garrisonroo, SidP, Kenyon, Tariqabjotu, Stephen, Velho, OwenX, Mindmatrix, TheNight-Fly, Ruud Koot, Jon Harald Søby, ZephyrAnycon, Teemu Leisti, BD2412, Nightscream, Koavf, Gmelli, Jweiss11, Tangotango, YAZA-SHI, Ggfevans, DirkvdM, FlaBot, Nihiltres, Fresheneesz, Skc0001, Alphachimp, Chobot, YurikBot, Borgx, Erachima, DTRY, Rick Nor-wood, Holycharly, TriGen, EEMIV, Bota47, Shadro, Mjhy0926, Allens, Infinity0, GrinBot~enwiki, DVD R W, Sardanaphalus, SmackBot, Aim Here, KocjoBot~enwiki, Thunderboltz, Stephensuleeman, WookieInHeat, Ieopo, The great kawa, Gilliam, Q1w2e3, Mhss, Psiphiorg, Bluebot, ViolinGirl, MalafayaBot, George Rodney Maruri Game, Octahedron80, DHN-bot~enwiki, Javelenok, Chlewbot, Mr.Z-man, Con-Man, Cybercobra, Jon Awbrey, RossF18, Byelf2007, The Ungovernable Force, SashatoBot, Nishkid64, Dbtfz, JoseREMY, IronGargoyle, Extremophile, Penagate, Comicist, Quaeler, Iridescent, K, Zarex, Van helsing, ChristineD, Neelix, Gregbard, Julian Mendez, Thisj!bot, LactoseTI, Marek69, Kborer, Noaa, AntiVandalBot, MaTT~enwiki, AaronY, IrishPete, Oliver Simon, BenMcLean, JAnDbot, Skomorokh, Agentnj, Hewinsj, GurchBot, Probios, Djradon, Kirrages, Rupes, VoABot II, Arno Matthias, Snowded, Oxford Dictionary, Illspirit, Van-ished user ty12kl89jq10, Cathalwoods, HemRaj Singh, Pbroks13, Pomte, Stjeanp, N4nojohn, J.delanoy, Trusilver, Shawn in Montreal, OAC, Tparameter, Jaxha, CompuChip, Heyitspeter, DorganBot, Vinsfan368, Lallallal, Jonwilliamsl, Pasixxxx, MARVINWAGNER, Rucha58, Leo-remy, TXiKiBoT, Technopat, A4bot, Msviolone, Philogo, Broadbot, Abdullais4u, Dprust, Andrewaskew, Graymornings, Lova Falk, Kusyadi, MCTales, Cnilep, Sfmammamia, SieBot, Paradoctor, Meldor, Mankar Camoran, Svick, DesolateReality, Mygerardromance, Escape Or-bit, Troy 07, Kenji000, De728631, ClueBot, R000t, Philosophy.dude, Bfeylia, Neverquick, Excirial, Rohit nit, GoldenGoose100, PaulK-incaidSmith, SpikeToronto, Ember of Light, GlassGhost, Thingg, Vanished User 1004, Zenwhat, XLinkBot, BodhisattvaBot, Kwork2, Ger-hardvalentin, Saeed.Veradi, WikiHead, Qwertytastic, Ewger, Addbot, CanadianLinuxUser, H92Bot, Glane23, GlobalSequence, Tide rolls, Lightbot, ScienceApe, لطفی Legobot, Luckas-bot, Yobot, 2D, Azcolvin429, AnomieBOT, Rubinbot, PresMan, Flewis, Prbclj25, Arthur-Bot, Xqbot, Doezxcty, S h i v a (Visnu), Lord Archer, Capricorn42, Xephras, Harkleither, Hjurgen, Ordning, Lord Bane, Ruby.red.roses, RowanEvans, RibotBOT, SEASONnmr, FrescoBot, Beclp, Pinethicket, Rushbugled13, Mohehab, JimRech, Jandalhandler, NeuroBells123, Keri, Humble Rat, Difu Wu, Whisky drinker, Badelmann, Tesseract2, DASHBot, EmausBot, Hedonistbot4000, Mo ainm, Tommy2010, Winner 42, TheGeomaster, JSquish, Fæ, Onecanwhile, MindShifts, Foreverlove642, Wayne Slam, Donner60, Tziemer991, Jimmynudes2, ClueBot NG, Drdoug5, Kimberleyporter, Fauzan, Jj1236, Tideflat, Amr.rs, Dictabeard, O.Koslowski, Masssly, Widr, Chilllls, Helpful Pixie Bot, Nichole773, Hallows AG, Wiki13, Luke13579, Richard84041, Ninjagoat, Sopidex, Dhruv-NJITWILL, Sgilmore10, Courtneysfoster, ChrisGualtieri, GoShow, Oligocene, ShangTsung87, Lugia2453, 93, MostlyListening, BreakfastJr, EMBViki, Strikingstar, VogelsangLorenzo, Ginsuloft, Mauriziogeri2013, Monkbot, KasparBot, Sorte Slyngel and Anonymous: 366

- **Inductive reasoning** Source: https://en.wikipedia.org/wiki/Inductive_reasoning?oldid=670953618 Contributors: AxelBoldt, The Cunctator,

The Anome, Ryguasu, DennisDaniels, Michael Hardy, Earth, Owl, Voidvector, BoNoMoJo (old), Jfitzg, Andres, Evercat, EdH, Desert-Steve, Timwi, Trontonian, Bemoeial, Ike9898, Wolfgang Kufner, Radiojon, Markhurd, Peregrine981, Banno, Nufy8, Nurg, Romanm, Ojigiri~enwiki, Mikiher, Tobias Bergemann, Filemon, Ancheta Wis, Giftlite, Zigger, Peruvianllama, Bovlb, Jason Quinn, Jackol, ELApro, Guppyfinsoup, Lucidish, Archer3, Discospinster, Freestylefrappe, Ivan Bajlo, Bender235, Kbh3rd, El C, Aaronbrick, David Crawshaw, Bobo192, I9Q79oL78KiL0QTFHgyc, Flammifer, Samohyl Jan, Yuckfoo, Mikeo, Recury, Nightstallion, Kinema, Kazvorpal, Kenyon, Hq3473, Velho, Mindmatrix, Kzollman, Ruud Koot, Alfakim, Andrea.gf, Rjwilmsi, Jweiss11, Strake, Bryan H Bell, Reinis, Matt Deres, Chris Pressey, Latka, RexNL, Fresheneesz, NotJackhorkheimer, Spencerk, King of Hearts, Chobot, Dresdnhope, YurikBot, RussBot, Gaius Cornelius, Grafen, Holycharly, SAE1962, 24ip, Pkearney, Roy Brumback, Bota47, Shadro, Tomisti, Sethery, Fram, Curpsbot-unicodify, Teply, Infinity0, Bo Jacoby, Jer ome, Sardanaphalus, SmackBot, RedHouse18, David Kernow, Rtc, McGeddon, Istvan, Eskimbot, Klokie, Gilliam, Duke Ganote, Ohnoitsjamie, Betacommand, Bluebot, Anthonzi, LaggedOnUser, DHN-bot~enwiki, DoctorStrangelove, Can't sleep, clown will eat me, Go For It, Avb, Edivorce, Mr.Z-man, Jmnbatista, Richard001, Kalexander, Jon Awbrey, Neshatian, Andeggs, Vina-iwbot~enwiki, Byelf2007, Jonrgrover, Normalityrelief, RichMorin, Antonielly, Aleenf1, Lukemcgrath, Grumpyyoungman01, Domokato, Levineps, Iridescent, K, Wjejskenewr, FleetCommand, CWY2190, Indigenius, El aprendelenguas, TMN, Gregbard, Slazenger, Peterdjones, Khromatikos, Gogo Dodo, Wikipediarules2221, Miguel de Servet, Letranova, Gacggt, Ucanlookitup, Second Quantization, Danny Reese, Defeatedfear, Fotomatt, AntiVandalBot, Luna Santin, Minhtung91, Spencer, Salgueiro~enwiki, JAnDbot, Davewho2, Dmar198, Coffee2theorems, Magioladitis, Bongwarrior, Equinexus, Hasek is the best, Arno Matthias, Farquaadhnchmn, DAGwyn, Snowded, Moopiefoof, Cathalwoods, MetsBot, Chrisdone, WLU, Mommyzbrat, STBot, Dionysiaca, Pomte, TheSeven, OttoMäkelä, LordAnubisBOT, Mahewa, Touisiau, Chiswick Chap, Heyitspeter, Pianoman55~enwiki, MetsFan76, STBotD, Andy Marchbanks, Straw Cat, Zach425, VolkovBot, Thewolf37, Pasixxxx, Hotfeba, Shinju, Jimmaths, Tiktuk, Philip Trueman, Deleet, Katoa, Jazzwick, Philogo, Abdullais4u, Jackfork, PDFbot, Anarchangel, Jor344, Shifter95, Cnilep, Harmonicemundi, PhysPhD, Jammycaketin, AlleborgoBot, Newbyguesses, Dwandelt, Matthew Yeager, Mark Klamberg, Flyer22, Bobklahn, Oxymoron83, Vanished user oij8h435jweih3, MiNombreDeGuerra, Bagatelle, Sunrise, Linkboyz, Melcombe, Oneforlogic, ClueBot, Farras Octara, Eric Schoettle, Niceguyedc, Vandalometer, Rbakels, Excirial, Jusdafax, Kikilamb, Estirabot, ChrisKalt, Hazzzzzz12, Lx 121, XLinkBot, Fastily, Gerhardvalentin, Tegiap, Saeed.Veradi, Skarebo, WikHead, Kwjbot, Kbdankbot, Tayste, Addbot, Tanhabot, Jtradke, Numbo3-bot, Tide rolls, ScienceApe, KUSSOMAK, Legobot, Luckas-bot, Oilstone, THEN WHO WAS PHONE?, AnomieBOT, Doingmorestuffonline, Vanakaris, Bob Burkhardt, Parthian Scribe, Xqbot, Lord Bane, Hanberke, A157247, F-22 Raptored, Omnipaedista, Ribot-BOT, Delbertpeach, Alialiac, FieldOperative, Paine Ellsworth, SBA1870, Machine Elf 1735, Pinethicket, Kiefer.Wolfowitz, Mavit0, A8UDI, Cleon7177, NeuroBells123, Gamewizard71, TobeBot, Jonkerz, Miracle Pen, Dbmikus, Hyperbytev2, Ripchip Bot, Elspru, NerdyScienceDude, George Richard Leeming, EmausBot, Elanguescence, Grjoni88, Logical Cowboy, Gfoley4, RenamedUser01302013, Mo ainm, ZéroBot, Leminh91, CanonLawJunkie, Oncenawhile, Wagino 20100516, Erianna, EricWesBrown, L Kensington, Just granpa, 28bot, ClueBot NG, Gareth Griffith-Jones, MelbourneStar, Ek65, Millermk, Schicagos, Tsunamicharlie, Albertttt, Thepigdog, Masssly, Widr, Helpful Pixie Bot, HMSSolent, Curb Chain, BG19bot, Wiki13, MusikAnimal, Jander80, Wandwiki, Blue Mist 1, Will.Oliver, Trailspark, RichardMills65, Ctasa221, Fangli997376557, ChrisGualtieri, Lhu720, Hagrid da fifth, Watchpup32, Neurocitizen, Oligocene, Moonstroller-2, Jochen Burghardt, M strat17, 90b56587, Reid12345, Londomollari42, EMBViki, Cauzality, Aubreybardo, Liz, Logicman2, Hoffoholi, Superploro, Temprack5446, Heller-rrr, Isambard Kingdom, Pretty Panther 26 and Anonymous: 398

- **Empirical evidence** *Source:* https://en.wikipedia.org/wiki/Empirical_evidence?oldid=670433911 *Contributors:* Stevertigo, Michael Hardy, Eurleif, Glenn, Kaihsu, Raven in Orbit, Markhurd, Martinphi, Robbot, ShaunMacPherson, Jason Quinn, Taak, Just Another Dan, Chowbok, Piotrus, Mvuijlst, Jameshowison, Mdd, Gary, Atlant, Joepzander, Kenyon, PeregrineAY, BD2412, Michaelbluejay, Jrtayloriv, Phoenix2~enwiki, VolatileChemical, RL0919, Scolaire, SmackBot, DCDuring, Jagged 85, Thumperward, Normxxx, Memming, Hammer1980, Madbull, Tktktk, Bjankuloski06en~enwiki, A. Parrot, Freder1ck, Khanssen, Levineps, K, Ncosmob, Danneckar, WeggeBot, Apterygial, Gregbard, Travelbird, ST47, TheM62Manchester, Ruaraidh-dobson, Runch, ThomasPusch, AntiVandalBot, Ingolfson, Ioeth, JAnDbot, Cgingold, Astragale, Geboy, Removepam, STBot, J.delanoy, Filll, ChrisfromHouston, Tdadamemd, Mikael Häggström, Fainites, Legoland12342, Jeff G., DianaGaleM, Poopsmith21, Jmath666, Falcon8765, Rurik3, Newbyguesses, Nihil novi, Gerakibot, Sunrise, Melcombe, Twinsday, Elassint, ClueBot, Gorillasapiens, Deselliers, Carolyn 3300, Bala990, Skarebo, Addbot, Ronhjones, Quercus solaris, Jubeidono, Yobot, Fmrauch, AnomieBOT, Tryptofish, LlywelynII, Trabucogold, Wortafad, Apollo, Xqbot, LogoX, Mikejens, Srich32977, Al Azif, GrouchoBot, Omnipaedista, Speednat, Trafford09, Pinethicket, Edderso, TobeBot, Ale And Quail, Dark Lord of the Sith, Theo10011, Reaper Eternal, GrindtXX, Wayne Slam, Donner60, Atlantictire, ClueBot NG, Mcgee123, Kathryn Whitney, JordanSeymore, Wbm1058, Snow Rise, Maestro814, Frosty, Malerooster, Mark viking, Theemathas, Aubreybardo, Saectar, Mohammad sohag khan, Loraof and Anonymous: 113

- **Mathematical induction** *Source:* https://en.wikipedia.org/wiki/Mathematical_induction?oldid=670793907 *Contributors:* AxelBoldt, Bryan Derksen, Zundark, Tarquin, Jan Hidders, XJaM, Mjb, Youandme, Bdesham, Michael Hardy, Wshun, David Martland, Dominus, Kku, Meekohi, Sannse, TakuyaMurata, Ahoerstemeier, Snoyes, Александър, Charles Matthews, Dcoetzee, Sbloch, Dysprosia, Jitse Niesen, Selket, Markhurd, Furrykef, Sabbut, Bloodshedder, .mau., Aenar, Robbot, Iwpg, Altenmann, Gandalf61, Bruceq, Aetheling, Ruakh, Tobias Bergemann, Tosha, Giftlite, Nickptar, Fintor, Peter Kwok, Kousu, PhotoBox, Lucidish, Shipmaster, Ericamick, Paul August, Bender235, Elwikipedista~enwiki, Zenohockey, CXI, Spoon!, Aplusbi, NetBot, Obradovic Goran, Polylerus, Jumbuck, Msh210, Arthena, Dirac1933, Bsadowski1, Oleg Alexandrov, Zntrip, Mindmatrix, StradivariusTV, Acone, Drostie, Ruud Koot, Mpatel, Sartak, M412k, Ryan Reich, Palica, Mandarax, Chun-hian, JIP, Rjwilmsi, SMC, FlaBot, Maitch, Mathbot, Chobot, YurikBot, Borgx, Hairy Dude, RussBot, Hyad, KSmrq, Debroglie, TEB728, Trovatore, Nick, Jstrater, Schmock, Pyroclastic, Froth, Gadget850, Bota47, Addps4cat, Mgnbar, Brisvegas, Jwissick, Reyk, SmackBot, RDBury, Selfworm, Melchoir, Pgk, Bomac, Pokipsy76, Jagged 85, Chris the speller, Fintler, JCSantos, Trebor, Nbarth, Nixeagle, ConMan, Radagast83, Acepectif, Lambiam, Wvbailey, J. Finkelstein, Pliny, Mets501, MissingNOOO, Dreftymac, Blackhawk charlie2003, Zero sharp, JRSpriggs, CRGreathouse, Wafulz, CBM, WeggeBot, Gregbard, Cydebot, Zahlentheorie, JFreeman, BillWeiss, Headbomb, Pcbene, AntiVandalBot, Vic226, Gramby, Sekky, Alphachimpbot, MER-C, Wisnuops, David Eppstein, DerHexer, SquidSK, Stephenchou0722, MartinBot, JCraw, Maurice Carbonaro, TomS TDotO, Paulecoyote, OttoMäkelä, It Is Me Here, LordAnubisBOT, Tparameter, Policron, David-CBryant, Omegamormegil, Fbarton, VolkovBot, AlnoktaBOT, Jimmaths, TXiKiBoT, Anonymous Dissident, Philogo, PaulTanenbaum, Gilisa, Pboulus, SieBot, Paradoctor, Alexsmail, MiNombreDeGuerra, Ellamosi, Ngriffeth, DesolateReality, Altzinn, Amahoney, AutoFire, Classicalecon, Gjakovit, ClueBot, Justin W Smith, The Thing That Should Not Be, Jdgilbey, Mathemajor, J8079s, DragonBot, Muro Bot, Tired time, SoxBot III, Dyhan81, Ost316, Brentsmith101, Kbdankbot, Addbot, DOI bot, Riyuky, CanadianLinuxUser, NjardarBot, MrVanBot, Chamal N, CarsracBot, LinkFA-Bot, Lightbot, Dminkovsky, Jarble, Yobot, Estudiarme, Jasonschock, AnomieBOT, 1exec1, MattTait, Citation bot, ArthurBot, TaySpace, Poetaris, DSisyphBot, J04n, RibotBOT, D'ohBot, Citation bot 1, Tkuvho, Kiefer.Wolfowitz, Unnachamois, Bozo the bear, Gabrielgmendonca, Mrhota, Vrenator, EdEveridge, Duoduoduo, Onel5969, Jowa fan, Whywhenwhohow, EmausBot, Vasanthloganathan, Slawekb, Ebrambot, Quondum, Noodleki, ClueBot NG, Moneysorter, Tideflat, B.dyck, Meingbg, Helpful Pixie Bot, DBigXray, FutureTrillionaire, Alpert1, CitationCleanerBot, Glacialfox, Justincheng12345-bot, Dexbot, Jochen Burghardt, I am One of Many, TI. Gracchus, Glasbys, Beneficii, Monkbot, FourViolas, Swashski, Dsacf1234567 and Anonymous: 230

- **Direct proof** *Source:* https://en.wikipedia.org/wiki/Direct_proof?oldid=636449332 *Contributors:* Michael Hardy, TakuyaMurata, Kaihsu, Korath, Nifboy, Peter Kwok, Andreas Kaufmann, Robertbowerman, Lachatdelarue, Gauge, Amerindianarts, Daranz, MarSch, GregAsche, Splintercellguy, Soltras, SmackBot, Luke Gustafson, Lambiam, Pliny, Bjankuloski06en~enwiki, 16@r, CBM, Gregbard, Cydebot, Thijs!bot, David Eppstein, STBot, VolkovBot, Jimmaths, Anonymous Dissident, Synthebot, Jdaloner, ClueBot, Estirabot, Mitch Ames, Addbot, Luckas-bot, La comadreja, GrouchoBot, RibotBOT, Erik9bot, MastiBot, Igor N Dimovski, Duoduoduo, EmausBot, ClueBot NG, Hoorayforturtles, Attempt4, Widr, Helpful Pixie Bot, BG19bot, CitationCleanerBot, Monkbot, Junaidjamshed, CS104G19, Unaizalakain, Rodebrecht and Anonymous: 14

- **Proof by contrapositive** *Source:* https://en.wikipedia.org/wiki/Proof_by_contrapositive?oldid=660854628 *Contributors:* Andyfugard, Centrx, Jason Quinn, Brianjd, BD2412, UsaSatsui, Doncram, Carabinieri, Geoffrey.landis, SmackBot, InverseHypercube, NickPenguin, Byelf2007, Gregbard, Cydebot, Daniel5Ko, Jimmaths, DesolateReality, PerryTachett, Badgernet, Alexander.mitsos, خ‌الد حسّن, Yobot, Erik9bot, BertSeghers, Jackofhats, Wcherowi, Helpful Pixie Bot, CitationCleanerBot, Anythingcouldhappen, Monkbot and Anonymous: 8

- **Proof by contradiction** *Source:* https://en.wikipedia.org/wiki/Proof_by_contradiction?oldid=665680179 *Contributors:* AxelBoldt, Magnus Manske, Lee Daniel Crocker, Vicki Rosenzweig, Mav, Zundark, The Anome, Tarquin, Larry Sanger, Andre Engels, Roadrunner, FvdP, B4hand, Patrick, Chas zzz brown, Michael Hardy, Oliver Pereira, DopefishJustin, Kidburla, Dominus, Dcljr, Skysmith, Andrewa, Scott, EdH, Ideyal, Revolver, Charles Matthews, Populus, Fibonacci, Leonariso, Robbot, Mohan ravichandran, Altenmann, Romanm, Chancemill, Sam Spade, Bkell, UtherSRG, Centrx, Giftlite, Barbara Shack, Everyking, Cortina, 20040302, Chrismear, Jason Quinn, Chameleon, Gubbubu, Toytoy, Michaelcarraher, Rdsmith4, Grossdomestic, Peter Kwok, Neale Monks, Lacrimosus, Hydrox, Roybb95~enwiki, Paul August, Brian0918, DrewRobinson, Cretog8, Dungodung, Man vyi, Jonsafari, Nsaa, Rd232, Burn, Bart133, Omphaloscope, Out180, Chamaeleon, Axeman89, Kazvorpal, Harvestdancer, Mindmatrix, Rodrigo Rocha~enwiki, Hdante, Graham87, BD2412, Chun-hian, Patrick Zanon, X1011, Mathbot, Markkbilbo, Mattman00000, Lemuel Gulliver, Glenn L, Chobot, Reetep, Algebraist, YurikBot, Severa, Zafiroblue05, Kerowren, Dkg11hu, Doctor Whom, TERdON, Bota47, Andersersej~enwiki, MagicOgre, SV Resolution, PurplePlatypus, Cmglee, Seanjacksontc, SmackBot, Pgk, BiT, Wje, Psiphiorg, Qwasty, ScottForschler, Jfsamper, Nbarth, Kobayen, Simpsons contributor, Aerobird, Ioscius, Mhym, Byelf2007, Lambiam, Dbtfz, Pliny, Loodog, Robofish, Mgiganteus1, Grumpyyoungman01, Xiaphias, Dr.K., Emx~enwiki, Ncosmob, Wafulz, CBM, Gregbard, Mattbuck, Cydebot, Shirulashem, MarcelLionheart, Epbr123, Kajisol, Pampas Cat, Flarity, CZeke, Hopiakuta, JAnDbot, Pedro, Looking-Glass, NeighborTotoro, Electriceel, TheBusiness, JuanPaBJ16, Neonguru, AltiusBimm, Colincbn, Pmbcomm, Kesal, DorganBot, VolkovBot, Jimmaths, TXiKiBoT, Liko81, Melsaran, Synthebot, GlassFET, Katzmik, Arkwatem, The Evil Spartan, Luciengav, Marc van Leeuwen, SilvonenBot, NjardarBot, MrVanBot, CarsracBot, Nikie42, AgadaUrbanit, Tide rolls, Alexander.mitsos, Legobot, Yobot, Cflm001, Majestic-chimp, AnomieBOT, OpenFuture, RavShimon, Dave3457, LegendFSL, Duoduoduo, PetroniusArb, DexDor, Jdl22, Bastian964, Makecat, Mehdi, ClueBot NG, Aflyhorse, Helpful Pixie Bot, Crh23, Vesta Zenobia, Vclam068, Jochen Burghardt, Password is DOB, Xin-Xin W., The Original Bob and Anonymous: 150

- **Constructive proof** *Source:* https://en.wikipedia.org/wiki/Constructive_proof?oldid=653981010 *Contributors:* Michael Hardy, Gandalf61, Tobias Bergemann, Giftlite, Macrakis, Peter Kwok, Txa, Pol098, Chenxlee, Jameshfisher, Gurch, Quuxplusone, Bgwhite, Jayme, Gaius Cornelius, SmackBot, Bluebot, Lambiam, Lim Wei Quan, CBM, Pierre de Lyon, Cydebot, Salgueiro~enwiki, RebelRobot, Dr Caligari, Jimmaths, Dmcq, SieBot, Classicalecon, Mahue, Leberbaum, Addbot, Unzerlegbarkeit, OlEnglish, Yobot, Obscuranym, AnomieBOT, Erel Segal, JRB-Europe, Twri, Noamz, FrescoBot, Steve2011, SMmoto, Thomassteinke, DARTH SIDIOUS 2, Slawekb, Chricho, ClueBot NG, Helpful Pixie Bot, Khonkhortisan, Jochen Burghardt, Acetotyce, LudicrousTripe and Anonymous: 25

- **Proof by exhaustion** *Source:* https://en.wikipedia.org/wiki/Proof_by_exhaustion?oldid=635976848 *Contributors:* Michael Hardy, Dcljr, Karada, Charles Matthews, Gandalf61, Ian Maxwell, Peter Kwok, PhotoBox, Jnestorius, DimaDorfman, Dfeldmann, Anthony Appleyard, StradivariusTV, Bubba73, Eubot, YurikBot, Hakeem.gadi, NYKevin, Cybercobra, Desmond71, Dbtfz, Robofish, Mets501, ShelfSkewed, Gregbard, Cydebot, Eurobas, Edward321, EnOreg, Classicalecon, Tautologist, Addbot, Yobot, Idealboat, EmausBot, Jumpythehat, Mywalnut, Mario Castelán Castro and Anonymous: 13

- **Probabilistic method** *Source:* https://en.wikipedia.org/wiki/Probabilistic_method?oldid=651459649 *Contributors:* Michael Hardy, Dominus, Charles Matthews, Viz, Kevinatilusa, Giftlite, Andris, Peter Kwok, Qutezuce, Paul August, Ryan Reich, Adking80, Mathbot, Maxal, Michael Slone, Shell Kinney, Ott2, Nealeyoung, Pierre de Lyon, Cydebot, Headbomb, Widefox, Lantonov, Nagy, Melcombe, ClueBot, Alexbot, M4gnum0n, Addbot, Dyaa, DemocraticLuntz, Erel Segal, Psdey1, Buenasdiaz, Miym, Charvest, Citation bot 1, Agrinshp and Anonymous: 18

- **Combinatorial proof** *Source:* https://en.wikipedia.org/wiki/Combinatorial_proof?oldid=573037719 *Contributors:* Michael Hardy, Mxn, Charles Matthews, Zarvok, Peter Kwok, Oleg Alexandrov, Will Orrick, Rjwilmsi, Alexb@cut-the-knot.com, Mhym, Bwsulliv, Stebulus, Sopoforic, Cydebot, MER-C, David Eppstein, Lantonov, PaulTanenbaum, Hffman, Arjayay, Marc van Leeuwen, Addbot, Citation bot 1, EmausBot, John of Reading, Joel B. Lewis, Helpful Pixie Bot, Brad7777 and Anonymous: 14

- **Statistical proof** *Source:* https://en.wikipedia.org/wiki/Statistical_proof?oldid=666752008 *Contributors:* Michael Hardy, Wtmitchell, Btyner, Rjwilmsi, Vegaswikian, Piet Delport, SmackBot, G716, AndrewHowse, Conquistador2k6, AlasdairBailey, Melcombe, Tautologist, Mild Bill Hiccup, SchreiberBike, Qwfp, Tayste, Ettrig, Yobot, Are you ready for IPv6?, J04n, Thompsma, Pollinosisss, RjwilmsiBot, Frietjes, Helpful Pixie Bot, Qetuth, Illia Connell and Anonymous: 3

- **Computer-assisted proof** *Source:* https://en.wikipedia.org/wiki/Computer-assisted_proof?oldid=657268339 *Contributors:* Poor Yorick, Dcoetzee, Furrykef, Stephan Schulz, Giftlite, Beland, Bender235, Rgdboer, John Vandenberg, Ultra megatron, Guthrie, Alai, Salix alba, Mallocks, Zvika, Bird of paradox, Nbarth, RyanEberhart, JonHarder, Germandemat, Acdx, Barabum, Skapur, CRGreathouse, Matthew Treder, Roccorossi, Hilgerdenaar, Oddity-, David Eppstein, MaD70, VanishedUserABC, Educres, Alik Kirillovich, Classicalecon, H.Marxen, Dthomsen8, MystBot, Addbot, Lightbot, Ptbotgourou, AnomieBOT, GnawnBot, Omnipaedista, Thosjleep, Mikrosam Akademija 3, Afteread, Helpful Pixie Bot, Hellachaz, Limit-theorem, Jmio17 and Anonymous: 25

- **Psychologism** *Source:* https://en.wikipedia.org/wiki/Psychologism?oldid=635449523 *Contributors:* Prosario 2000, Kku, Jm34harvey, Wclark, Chalst, Oop, Knucmo2, Keenan Pepper, Oleg Alexandrov, FlaBot, Pigman, Nicke L, M3taphysical, Tomisti, TransUtopian, Jules.LT, Bluebot,

JoshuaZ, Gregbard, Cydebot, Thijs!bot, William Knorpp, JAnDbot, Xeno, Pomte, Stdbrouw, DorganBot, VolkovBot, The Tetrast, Billinghurst, Jackbars, Myrvin, Hazzzzzz12, Addbot, SpBot, AnomieBOT, Dr Oldekop, GrouchoBot, Omnipaedista, Updatehelper, Bluszczokrzew, Emaus-Bot, ZéroBot, BG19bot, Gibbja, BattyBot, BreakfastJr, Rodney.k.b.parker and Anonymous: 11

- **Language of thought hypothesis** *Source:* https://en.wikipedia.org/wiki/Language_of_thought_hypothesis?oldid=648160731 *Contributors:* Edward, Cadr, Banno, Diberri, Eequor, Mporch, D6, Posiduck, Gary, Keenan Pepper, Velho, Ylem, Mandarax, Rjwilmsi, KYPark, Spencerk, Ncsaint, SmackBot, Smec, WikiPedant, OrphanBot, JBel, StN, Goodnightmush, Gregbard, Peterdjones, Quibik, Thrapper, Ninjakannon, Harborsparrow, Nono64, VanishedUserABC, Newbyguesses, Jojalozzo, Ddgromit, Krzysztofgajewski, EPadmirateur, Alexbot, Brews ohare, Lucky Bottlecap, Earcanal, Addbot, Redheylin, Yobot, MMPedwardg, Unara, FrescoBot, Abductive, Jemoore31688, Dmanzelmann, Give-nunion, Helpful Pixie Bot, 220 of Borg, Cgeggie, Jochen Burghardt, Aubreybardo, Philosopher of Mind and Anonymous: 28

- **Automated theorem proving** *Source:* https://en.wikipedia.org/wiki/Automated_theorem_proving?oldid=662891220 *Contributors:* AxelBoldt, The Anome, Taw, Dwheeler, Cwitty, Michael Hardy, JakeVortex, Karada, Rotem Dan, Charles Matthews, Dcoetzee, Dysprosia, Michaeln, Jimbreed, Stephan Schulz, TittoAssini, Wikibot, Lumingz, Tobias Bergemann, Ancheta Wis, Giftlite, Bfinn, Bobblewik, Beland, TheosThree, Tc~enwiki, Rich Farmbrough, Rama, Peter M Gerdes, AshtonBenson, Diego Moya, Nealcardwell, Krischik, Pontus, Alai, LunaticFringe, Oleg Alexandrov, Geoffgeoffgeoff3, Paul Haroun, Ruud Koot, JosefUrban, Graham87, Qwertyus, Grammarbot, Rjwilmsi, Tizio, MZMcBride, Brighterorange, Mathbot, Ysangkok, CarolGray, JYOuyang, Jrtayloriv, GreyCat, Nehalem, Bgwhite, Algebraist, Ksyrie, Grafen, Jp-bowen, Arthur Rubin, Saeed Jahed, Zmoboros, PhS, Nahaj, Zvika, Pintman, InverseHypercube, McGeddon, Bluebot, Clconway, MovGP0, Haberg, Slawekk, Akriasas, Jon Awbrey, Byelf2007, Ramykaram, Disavian, Michael Bednarek, Antonielly, Lancebledsoe, Loadmaster, JHalcomb, ILikeThings, CRGreathouse, CBM, Pgr94, Ezrakilty, Blaisorblade, Agent1, Magioladitis, Detla, Cic, David Eppstein, Epsilon0, Zacchiro, Laurusnobilis, Nattfodd, Uwe Bubeck, MaD70, BotKung, Brian Huffman, Synthebot, Newbyguesses, Arapajoe, DaYZman, Vanished user oij8h435jweih3, PaulBrinkley, BSoD, DainDwarf, Tatzelworm, Logperson, JohnAspinall, Eusebius, Adrianwn, Simon04, Dkf11, D.scain.farenzena, Tim32, Bracton, Ceilican, WikHead, Tassedethe, Lightbot, Jarble, Legobot, Adelpine, Linket, AnomieBOT, Citation bot, Doezxcty, FrescoBot, Wikinglouis, Iislucas, Kishmakov, Trappist the monk, Vincent Aravantinos, Jfmantis, BertSeghers, Mekeor, Chricho, Peskoj, Thetna123, Sonic7406, Arrandale, Frietjes, David Luckham, Taneltammet, BattyBot, ChrisGualtieri, Khazar2, Y256, Jochen Burghardt, Pimp slap the funk, Ragerdl, Monkbot, SiddMahen, NQ, Kyle1009 and Anonymous: 142

- **Mathematical fallacy** *Source:* https://en.wikipedia.org/wiki/Mathematical_fallacy?oldid=667987515 *Contributors:* AxelBoldt, Bryan Derksen, Zundark, Arvindn, Michael Hardy, Dominus, Eric119, Minesweeper, Ijon, LittleDan, UserGoogol, Schneelocke, Charles Matthews, Timwi, Dcoetzee, Dysprosia, Jitse Niesen, Wik, Wiwaxia, Fredrik, Altenmann, Gandalf61, Merovingian, Henrygb, Tobias Bergemann, Giftlite, Wolfkeeper, Paul Pogonyshev, Guanaco, Matt Crypto, Mdob, Chowbok, Bongbang, Starx, Peter Kwok, Gdabski, Gazpacho, TheJames, Paul August, ESkog, Lankiveil, Spoon!, Nandhp, JRM, Wood Thrush, I9Q79oL78KiL0QTFHgyc, Martinultima, Tsirel, JYolkowski, Anders Kaseorg, SurrealWarrior, Splat, Mikeo, Drbreznjev, AxeMan89, Feezo, StradivariusTV, Apokrif, Waldir, BD2412, Jshadias, Josh Parris, Eyu100, R.e.b., Tomtheman5, Tedd, Alexb@cut-the-knot.com, Mathbot, Celestianpower, King of Hearts, Sbrools, DVdm, X42bn6, RussBot, Ian-Manka, Gaius Cornelius, Pnrj, Cheeser1, Simxp, Syko, Brentt, KnightRider~enwiki, SmackBot, Incnis Mrsi, Melchoir, Rokfaith, Fulldecent, Bluebot, Thumperward, SchfiftyThree, Tavianator, RyanEberhart, Calbaer, Fuhghettaboutit, Pwjb, Turms, Louisng114, Henning Makholm, Ged UK, Byelf2007, Zchenyu, Lambiam, Polihale, Omnedon, Illythr, Cstella23, Kpengboy, Mets501, Hyperwiz, Dr.K., AlsatianRain, JR-Spriggs, George100, Whyareall, CRGreathouse, CBM, JPadron, Cydebot, Reywas92, WillowW, MC10, Steel, Crossmr, Carifio24, Kacie Jane, Odie5533, Krzysiu Jarzyna, Englishnerd, Yurell, Kilva, Pallas44, Jojan, AntiVandalBot, Seaphoto, Joe Schmedley, MarvinC2, Dylan Lake, Husond, Oxinabox, MER-C, Boleslaw, Drhlajos, Some Guy123, Timanderso, Albmont, Email4mobile, Catgut, MetsBot, Error792, Cpl Syx, Dravick, Patstuart, Connor Behan, Ztobor, MartinBot, Ariel., Xoran99, Arjun01, Comperr, Anaxial, J.delanoy, AstroHurricane001, Uncle Dick, Laurusnobilis, Paidgenius, GEWilker, Soccersabo, Useight, TWiStErRob, RJASE1, Kimandy, Science4sail, Indubitably, Nousernamesleft, Wannger27, Anonymous Dissident, Amahdy, Kmhkmh, Geoffreyfishing, Secretss, Dmcq, Sue Rangell, Misha Mullov-Abbado, Oboeboy, Paradoctor, Phe-bot, Keilana, Happysailor, Dragnmn, Taemyr, 0rrAvenger, Kudret abi, 🔲🔲🔲🔲~enwiki, Kortaggio, Tuntable, Clue-Bot, EoGuy, Mild Bill Hiccup, Xenon54, Oxnard27, Doloco, Fletcher17, Lartoven, Ykhwong, H.Marxen, Djk3, XLinkBot, Marc van Leeuwen, Fastily, Pichpich, Tongrongtian, Gwandoya, Gerhardvalentin, Charles Sturm, AlexFekken, Luca Antonelli, Nickolai kazímir, Addbot, Joe-Moron2000, 067012732s, CanadianLinuxUser, Fluffernutter, Favonian, Barak Sh, 3wertbob7, Calculuslover, Tide rolls, LNKapustin, Yobot, TaBOT-zerem, Timeroot, Spenalzo, AnomieBOT, Pkukiss, Jim1138, AdjustShift, Terminatore, Georgepowell2008, Xqbot, TechBot, The Evil IP address, Point-set topologist, POTUS270, Jetpackboy14, Pottersson, Pinethicket, Number Googol, Patwotrik, Tcnuk, Jujutacular, Barras, Double sharp, Trappist the monk, Niketmalik, Phatency, Le Docteur, Thewriter006, Tbhotch, Sideways711, Whisky drinker, Martianpackets, Mr. Anon515, EmausBot, John of Reading, Slawekb, Anoop.dixith, Derekleungtszhei, Quondum, L Kensington, JonRichfield, ClueBot NG, Rtucker913, Helpful Pixie Bot, Hguy, BG19bot, Hawkwindeb, Mocky3497, Nicolae-boicu, Avengingbandit, Rupert'sscribe, Saung Tadashi, Lugia2453, Mmitchell10, Yehianumb, Gtklocker, Bilorv, That kiwi guy, Mario Castelán Castro, Stishuk.hf, PiotrGrochowski000, Gov vj, Abhishekx7 and Anonymous: 286

- **List of incomplete proofs** *Source:* https://en.wikipedia.org/wiki/List_of_incomplete_proofs?oldid=662738086 *Contributors:* AxelBoldt, Zundark, Michael Hardy, Dominus, Schneelocke, Giftlite, Jason Quinn, Paul August, Kzollman, Rjwilmsi, R.e.b., Spacepotato, Geraschenko, SmackBot, RDBury, Espresso Addict, PrimeHunter, Scwlong, Tamfang, Makyen, Harlekeyn, JRSpriggs, CmdrObot, Cydebot, MC10, Blicher, Quibik, BetacommandBot, Headbomb, RobHar, Hermel, Fabrictramp, David Eppstein, DGG, Hasteur, Cold Phoenix, Pichpich, Yobot, Kilom691, Citation bot 1, Kiefer.Wolfowitz, Xnn, RjwilmsiBot, Helpful Pixie Bot, K9re11, SoSivr and Anonymous: 7

- **List of long mathematical proofs** *Source:* https://en.wikipedia.org/wiki/List_of_long_mathematical_proofs?oldid=665978773 *Contributors:* Edward, TakuyaMurata, Charles Matthews, Dfeldmann, Apokrif, Rjwilmsi, R.e.b., Drone5, Myasuda, Cydebot, David Eppstein, Kope, Olson-ist, R'n'B, Yobot, The Evil IP address, Trappist the monk, LucasBrown, GoingBatty, Brandmeister, EefeG0hi, K9re11, A.lisitsa and Anonymous: 3

- **List of mathematical proofs** *Source:* https://en.wikipedia.org/wiki/List_of_mathematical_proofs?oldid=651793994 *Contributors:* Manning Bartlett, Edward, Michael Hardy, Dominus, Revolver, Pfortuny, Marc Venot, Tosha, Giftlite, Dbenbenn, Dissident, Gro-Tsen, Golbez, Zarvok, Peter Kwok, Shahab, Paul August, ZeroOne, ABCD, Oleg Alexandrov, Jacobolus, Salix alba, R.e.b., Mathbot, YurikBot, IanManka, Grubber, Buster79, Figaro, Googl, Zvika, Silly rabbit, Syrcatbot, Cydebot, Neko244, Weixifan, Leon math, JohnBlackburne, Jimmaths, Synthebot, Dmcq, OlEnglish, RJGray, Set theorist, Wbm1058, Joemkhan and Anonymous: 9

- **Proof by intimidation** *Source:* https://en.wikipedia.org/wiki/Proof_by_intimidation?oldid=651695706 *Contributors:* The Anome, Michael Hardy, Dominus, Altenmann, Rheun, Nerdfiles, Florian Blaschke, Linas, Rjwilmsi, Finell, SmackBot, Bill3000, Lambiam, Antonielly, Harej bot, Cydebot, MC10, EdJohnston, Magioladitis, Bearian, JackSchmidt, Nnemo, Pichpich, Addbot, DOI bot, Citation bot, Xqbot, Toa Nidhiki05, DrilBot, PowerWiki112233, Lotje, Scientific29, KlappCK and Anonymous: 17

- **Termination analysis** *Source:* https://en.wikipedia.org/wiki/Termination_analysis?oldid=651793948 *Contributors:* Palmcluster, Mdd, Diego Moya, Ruud Koot, BD2412, Robert A West, Robertbyrne, Garion96, SmackBot, Derek farn, Wvbailey, Jafet, Konstantin.Solomatov, Bazzargh, AnAj, Faizhaider, Yobot, AnomieBOT, Gilo1969, J04n, ClueBot NG, Braincricket, CarrieVS, Jochen Burghardt, Nomoteretes and Anonymous: 8

- **Mathematical beauty** *Source:* https://en.wikipedia.org/wiki/Mathematical_beauty?oldid=665571253 *Contributors:* Tarquin, Miguel~enwiki, Michael Hardy, Nixdorf, MartinHarper, GTBacchus, Karada, Ahoerstemeier, DavidWBrooks, Jimfbleak, Angela, Nikai, Nikola Smolenski, Revolver, Dysprosia, KRS, Ann O'nyme, Jmartinezot, Jose Ramos, Bevo, Aleph4, Gandalf61, Hadal, Randomness~enwiki, Ancheta Wis, Giftlite, Dbenbenn, Lupin, Bfinn, WHEELER, Dav4is, JRR Trollkien, Andycjp, Alexf, Elroch, Sam Hocevar, Gscshoyru, ChaTo, Thorwald, Paul August, Billymac00, C S, Andrewbadr, IonNerd, Orimosenzon, Critical, Linas, Mindmatrix, LOL, Ruud Koot, Tedneeman, Ryan Reich, Lawrence King, Grammarbot, Hack-Man, Salix alba, Slac, R.e.b., Nihiltres, Jersey Devil, DVdm, YurikBot, Wolfmankurd, Wikinick~enwiki, NawlinWiki, Arichnad, Raven4x4x, EEMIV, Sandstein, Cullinane, Gesslein, Allens, SmackBot, Stepa, PeterSymonds, Kithburd, SMP, Silly rabbit, Nbarth, DHN-bot~enwiki, Scalene, Rrelf, Armend, Jon Awbrey, DMacks, Xiutwel, Lambiam, Howdoesthiswo, Amenzix, Random-Critic, Waggers, Mets501, E-Kartoffel, BranStark, Aeternus, Sakurambo, CRGreathouse, Wafulz, A civilian, WeggeBot, Gregbard, Nauticashades, Tsenapathy, MC10, M a s, Int3gr4te, Thijs!bot, Kilva, Liquid-aim-bot, Mhaitham.shammaa, Narssarssuaq, HarmonicFeather, Xeno, David Eppstein, Ineffable3000, Maurice Carbonaro, Nigholith, Rnest2002, DadaNeem, Ogranut, Funandtrvl, Hammersoft, Chitownmack, SteveStrummer, Broadbot, Lambyte, Psyche825, Everything counts, Mouse is back, Billinghurst, Adam.J.W.C., Euryalus, Anchor Link Bot, Tautologist, ClueBot, KarenSutherland, MonoBot, XLinkBot, Addbot, Idbelange, Guffydrawers, Tassedethe, Yobot, Fleabox, AnomieBOT, Citation bot, ProtectionTaggingBot, Kurosuke88, Ron Aharoni, Citation bot 1, Daclyff, EdEveridge, PPdd, WikitanvirBot, Kiatdd, ZéroBot, Thewhyman, Mixedberries17, ZeroCool4ta, Fuzzy artist, Miegoreng, Rmashhadi, ClueBot NG, Frietjes, Delusion23, Reify-tech, Luckimg, Helpful Pixie Bot, BG19bot, Brad7777, Pankaj Jyoti Mahanta, Mathbeauty, Ashorocetus, Nigellwh, Lamaballa, LZNQBD, Hampton11235, Blue-Continent and Anonymous: 116

- **Proofs from THE BOOK** *Source:* https://en.wikipedia.org/wiki/Proofs_from_THE_BOOK?oldid=652961256 *Contributors:* Kpjas, Michael Hardy, Markhurd, Gandalf61, Giftlite, Marcika, DragonflySixtyseven, Mani1, PaulHanson, Jeff3000, Salix alba, Bubba73, Jaraalbe, Cecil-Ward, SmackBot, Wdvorak, E-Kartoffel, CBM, Cydebot, Skomorokh, Matthew Komorowski, David Eppstein, TXiKiBoT, PaulTanenbaum, Lamro, AlleborgoBot, Fairstar, Palnot, Addbot, Luckas-bot, Citation bot, Twri, Citation bot 1, Peepo36, Andreschulz, Darvii and Anonymous: 16

36.2.2 Images

- **File:4CT_Non-Counterexample_1.svg** *Source:* https://upload.wikimedia.org/wikipedia/commons/a/a6/4CT_Non-Counterexample_1.svg *License:* Public domain *Contributors:* Based on a this raster image by Dmharvey on en.wikipedia. *Original artist:* Inductiveload

- **File:ANL-E_aerial_22037k4.jpg** *Source:* https://upload.wikimedia.org/wikipedia/commons/0/07/ANL-E_aerial_22037k4.jpg *License:* Public domain *Contributors:* Argonne National Laboratory photo [1] *Original artist:* Argonne National Laboratory

- **File:Ambox_important.svg** *Source:* https://upload.wikimedia.org/wikipedia/commons/b/b4/Ambox_important.svg *License:* Public domain *Contributors:* Own work, based off of Image:Ambox scales.svg *Original artist:* Dsmurat (talk · contribs)

- **File:Aristotle_Altemps_Inv8575.jpg** *Source:* https://upload.wikimedia.org/wikipedia/commons/a/ae/Aristotle_Altemps_Inv8575.jpg *License:* Public domain *Contributors:* Jastrow (2006) *Original artist:* Copy of Lysippus

- **File:Arithmetic_symbols.svg** *Source:* https://upload.wikimedia.org/wikipedia/commons/a/a3/Arithmetic_symbols.svg *License:* Public domain *Contributors:* Own work *Original artist:* This vector image was created with Inkscape by Elembis, and then manually replaced.

- **File:Brain.png** *Source:* https://upload.wikimedia.org/wikipedia/commons/7/73/Nicolas_P._Rougier%27s_rendering_of_the_human_brain.png *License:* GPL *Contributors:* http://www.loria.fr/~{}rougier *Original artist:* Nicolas Rougier

- **File:CollatzFractal.png** *Source:* https://upload.wikimedia.org/wikipedia/commons/1/1c/CollatzFractal.png *License:* Public domain *Contributors:* English wikipedia *Original artist:* Pokipsy76

- **File:Commons-logo.svg** *Source:* https://upload.wikimedia.org/wikipedia/en/4/4a/Commons-logo.svg *License:* ? *Contributors:* ? *Original artist:* ?

- **File:Compound_of_five_cubes.png** *Source:* https://upload.wikimedia.org/wikipedia/commons/0/03/Compound_of_five_cubes.png *License:* Attribution *Contributors:* Transferred from en.wikipedia to Commons by Sreejithk2000 using CommonsHelper. *Original artist:* Tomruen at English Wikipedia

- **File:Diagram_of_Pythagoras_Theorem_simplified.png** *Source:* https://upload.wikimedia.org/wikipedia/commons/6/6e/Diagram_of_Pythagoras_Theorem_simplified.png *License:* CC BY-SA 4.0 *Contributors:* {{<a href='//commons.wikimedia.org/wiki/File:Diagram_of_Pythagoras_Theorem.png'>Existing wikipedia diagram - cropped</a>}} *Original artist:* Unaizalakain

- **File:Dominoeffect.png** *Source:* https://upload.wikimedia.org/wikipedia/commons/9/92/Dominoeffect.png *License:* CC-BY-SA-3.0 *Contributors:* ? *Original artist:* ?

- **File:Edit-clear.svg** *Source:* https://upload.wikimedia.org/wikipedia/en/f/f2/Edit-clear.svg *License:* Public domain *Contributors:* The *Tango! Desktop Project*. *Original artist:*
The people from the Tango! project. And according to the meta-data in the file, specifically: "Andreas Nilsson, and Jakub Steiner (although minimally)."

www.ingramcontent.com/pod-product-compliance
Lightning Source LLC
Chambersburg PA
CBHW072302200526
45168CB00014B/213